THE BASICS OF RELATIVITY

THE BASICS OF RELATIVITY

DR. ALIMUDDIN KHAN

ANMOL PUBLICATIONS PVT. LTD.
NEW DELHI - 110 002 (INDIA)

ANMOL PUBLICATIONS PVT. LTD.

H.O.: 4374/4B, Ansari Road, Darya Ganj,
New Delhi-110 002 (India)
Ph.: 23278000, 23261597

B.O.: No. 1015, Ist Main Road, BSK IIIrd Stage
IIIrd Phase, IIIrd Block
Bangalore - 560 085 (India)
Visit us at: www.anmolpublications.com

The Basics of Relativity

ISBN 978-81-261-3383-3

PRINTED IN INDIA

Printed at Mehra Offset Press, Delhi.

Contents

Preface

General relativity is the most beautiful physical theory ever invented. It describes one of the most pervasive features of the world we experience – gravitation – in terms of an elegant mathematical structure – the differential geometry of curved space-time – leading to unambiguous predictions that have received spectacular experimental confirmation. Consequences of general relativity, from the big bang to black holes, are often what get young people first interested in physics, and it is an unalloyed joy to finally reach the point in one's studies where these phenomena may be understood at a rigorous quantitative level. If you are contemplating reading this book, that point is here.

In recent decades, general relativity (GR) has become an integral and indispensable part of modern physics. For a long time after it was proposed by Einstein in 1916, GR was counted as a shining achievement that lay somewhat outside the mainstream of interesting research. Increasingly, however, contemporary students in a variety of specialties are finding it necessary to study Einstein's theory.

In addition to being an active research area in its own right, GR is part of the standard syllabus for anyone interested in astrophysics, cosmology, string theory, and even particle physics. This is not to slight the more pragmatic uses of GR, including the workings of the Global Positioning System (GPS) satellite network. There is no shortage of books on GR, and many of them are excellent.

Indeed, approximately thirty years ago witnessed the appearance of no fewer than three books in the subject, each of which has become a classic in its own right: those by Weinberg (1972), Misner, Thorne, and Wheeler (1973), and Hawking and Ellis (1975). Each of these books is suffused with a strongly-held point of view advocated by the authors.

This has led to a love-hate relationship between these works and their readers; in each case, it takes little effort to find students who will declare them to be the best textbook ever written, or other students who find them completely unpalatable. For the individuals in question, these judgments may very well be correct; there are many different ways to approach this subject.

Author

Chapter 1

Galilean Era and Newton Science

GALILEO SCIENCE

MECHANICS

Most of Galileo's investigations in physics had to do with the motion of bodies; these investigations lead him to the modern description of motion in terms of position and time. He realized that two important quantities that describe the motion of all bodies are velocity (which determines how position changes with time) and acceleration (which determines how velocity changes with time)

Two Important Defitnitions

Velocity: The rate of change of position (how position changes with time)

Acceleration: The rate of change of velocity (how velocity changes with time)

The motion of falling bodies: Galileo realized, even during his earliest studies (published in his book On motion) that the speed of a falling body is independent of its weight. He argued as follows: suppose, as Aristotle did, that the manner in which a body falls does depend on it weight (or on some other quality, such as its "fiery" or "earthy" character), then, for example, a two pound rock should fall faster than a one pound rock. But if we take a two pound rock, split it in half and join the halves by a light string then one the one hand this contraption should fall as fast as a two pound rock, but on the other hand it should fall as fast as a one-pound rock . Since any object should have a definite speed as it falls, this argument shows that the Aristotle's assumption that the speed of falling bodies is determined by their weight is inconsistent; it is simply wrong. Two bodies released from a given height will reach the ground (in general) at different times not because they have different "earthliness" and "fiery" characteristics, but merely because they are affected by air friction differently. If the experiment is tried in vacuum any two objects when released from a given height, will reach the ground simultaneously (this was verified by the Apollo astronauts on the Moon using a feather and a wrench).

This result is peculiar to gravity, other forces do not beahve like this at all. For example, if you kick two objects (thus applying a force to them) the heavier one will move more slowly than the lighter one. In contrast, objects being affected by gravity (and starting with the same speed) will have the same speed at all times. This unique property of gravity was one of the motivations for Einstein's general theory of relativity.

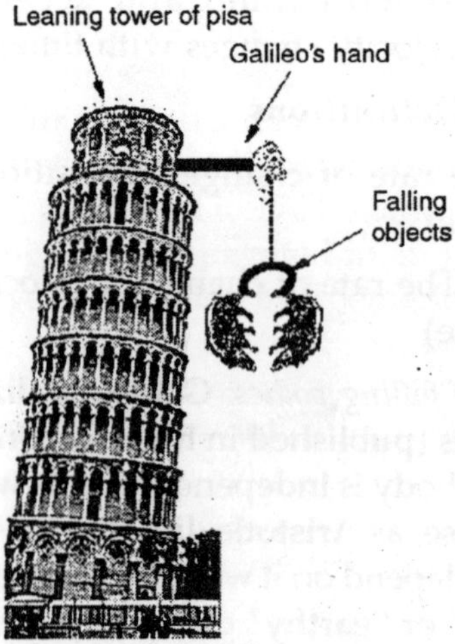

Fig. Galileo Experiment

Also in his investigations of falling bodies Galileo determined that the acceleration of these bodies is constant.

He demonstrated that an object released from a height starts with zero velocity and increases its speed with time (before him it was thought that bodies when released acquire instantaneously a velocity which remained constant but was larger the heavier the object was). Experimenting with inclined planes, and measuring a ball's positions after equal time intervals Galileo discovered the mathematical expression of the law of falling bodies: the distance increases as the square of the time.

THE MOTION OF PROJECTILES

Galileo also considered the motion of projectiles. He showed that their motion can be decomposed in a motion along a vertical and horizontal directions. Thus if a ball is thrown horizontally (and air friction is ignored) it will move in the horizontal direction with constant speed; in the vertical direction it will experience the pull of gravity and will undergo free fall. The use of this can be illustrated by the following situation. Suppose a ball is let fall from a height h and is found to take t seconds to reach the ground. Now suppose that the ball is instead thrown horizontally with speed v, what distance will it cover? The answer is v t because the ball, even though it is moving horizontally, in the vertical direction is still freely falling: the two motions are completely independent. This, of course, was of great use in warfare.

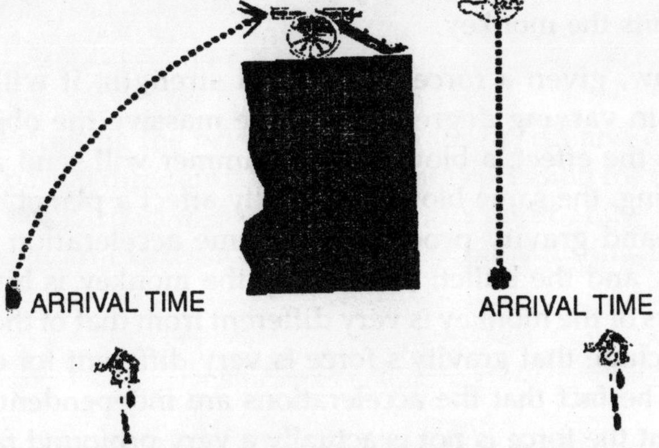

Fig. Galileo Experiment

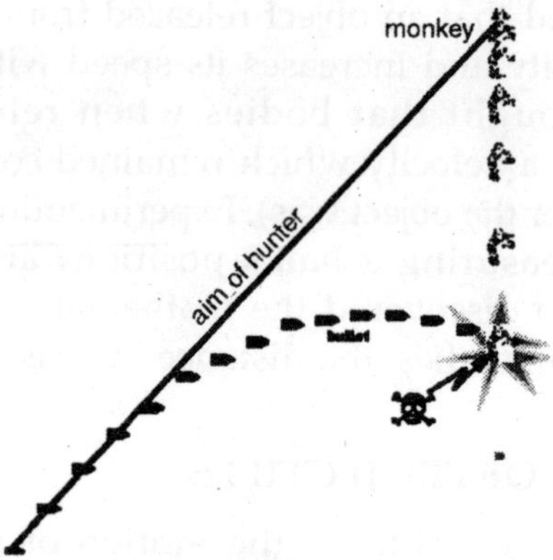

Fig. Galileo Law

The reason is the following: if there were no forces the bullet would go in a straight line (as indicated by the dotted line in the figure) and the monkey would not fall. So the bullet would hit the monkey. Now, since we have a force acting on the system (gravity) the monkey will not stay at rest but will accelerate downward. But precisely the same force acts on the bullet in precisely the same way, hence the bullet will not go in a straight line but will follow the curve indicated in the figure. The deviation from their force-free motions (rest for the monkey, straight line for the bullet) are produced by a force which generates the same acceleration in both objects, hence these deviations are precisely matched in such a way that the bullet hits the monkey.

Now, given a force of constant strength, it will affect bodies in varying degrees; the more massive the object the smaller the effect: a blow from a hammer will send a small ball flying, the same blow will hardly affect a planet. On the other hand gravity produces the same acceleration on the monkey and the bullet; that is why the monkey is hit. Since the mass of the monkey is very different from that of the bullet we conclude that gravity's force is very different for each of them. The fact that the accelerations are independent of the mass but the force is not is actually a very profound fact: the whole of general relativity is based on it.

GALILEAN PRINCIPLE OF RELATIVITY

Other than his numerous experiments on acceleration and the motion of objects, Galileo was also very observant, and he probably liked to ponder deeply about the things he see.

For example, he wondered whether it is possible for sailors in a boat to tell whether the ship is actually moving or not, without looking at the surrounding seascape and shoreline. To investigate this, he devised a simple experiment: the sailor will drop a ball when the boat is docking in bay, and when the boat is moving with uniform velocity in the open sea.

Of course, in practice there will be complications: depending on the weather the sea can be quite rough, and the light boats in Galileo's time will be heaved up and down by the choppy waters, and the result of the experiment will be quite different from that performed in bay, where the waters are calmer.

So Galileo imagined an ideal windless day where the waters in the open sea is calm and smooth, as would the waters in the bay. In such an ideal situation, Galileo argued that there will no difference in the outcomes of the experiments, be it done in the bay or out in the open sea. Now, if the sailor is locked in the cabin with no portholes to look out from, what would he conclude from the experiment?

What else?

"The boat is not moving, when I did my two experiments."

said the sailor, even though the boat is indeed moving with uniform velocity in the second case.

Actually, this was not what he said exactly. However, in his book Dialogue Concerning the Two Chief World Systems Galileo wrote a parable on a dialogue between two of the three chief characters, SALVATIUS and SAGREDUS in which SALVATIUS painted a hypothetical situation in which one lock himself or herself in the main cabin of a large ship together with a zoo, literally, and another person do the same in an identical ship. One of this ship is moving with uniform velocity on a smooth sea and the other one at rest.

In the parable, SALVATIUS's conclusion is that if both persons make careful observations on the two zoos on the moving ship and the stationary ship, then none of their observations about the motion of butterflies and bees flying and people throwing balls at each other, will differ from each other.

This leads to the Galilean Principle of Relativity, which was first stated as

The mechanical laws of physics are the same for every observer moving uniformly with constant speed in a straight line.

Ah... but this seems to imply that there must an observer who is absolutely at rest, as postulated by Aristotle, otherwise how do we know that any other observer is moving with a uniform velocity?

But what the Galilean Principle of Relativity did is precisely to remove the need for such an absolutely resting observer. We may argue for this by referring to Galileo's ship parable, in which the uniformly moving observer sees the same physics as the stationary observer. Therefore, if there is any absolutely resting observer, he will share the same physics will other uniformly moving observers in the sense all physical experiments performed in their closed main cabin will obtain the same results, relative to coordinates and times measured in this closed cabin.

However, if all physical measurements coincide, what serves to distinguish the absolutely resting observer's cabin from the uniformly moving observer's cabin? To tell the truth, nothing! If there is something that allows us to tell them apart, then clearly the two cabins are not equivalent physically, implying there must be at least one physical experiment whose results are different for the absolutely resting observer and the uniformly moving observer. But this would mean that Galileo's conclusion about the equivalence of physics is wrong!

Therefore, we have two choices:

Galilean Principle of Relativity TRUE = No absolutely resting observer.

Galilean Principle of Relativity FALSE = Absolutely resting observer possible.

Since then many empirical tests on the validity of this Principle has been carried out, and it is now clear to us that, at low velocities (much less than the speed of light) and not near strongly gravitating bodies, Galilean's Principle of Relativity has to be accepted as valid, and it is on this principle that Newton's Laws of Motions are based upon.

NEWTON'S LAWS

Newtonian mechanics, including the aspects developed by Galileo and others, was at least as revolutionary as Einstein's theory. In the West at least, ideas about mechanics had previously been dominated by the writings of Aristotle.

One of the big differences was this: for Aristotle, the 'natural' state of matter was at rest (with respect to the Earth). For Galileo and Newton, the 'natural' state of matter could be said to be constant velocity. For Aristotle, there is no need to ask why common objects come to a stop: that is their natural state. Rather, there is a need to ask why some objects (eg planets) keep moving. Now look at Newton's laws.

Newton's first law: If no force acts on a body, then it travels in a straight line at constant speed. ie its velocity is constant and its acceleration is zero.

Every body perseveres in its state of rest, or of uniform motion in a right line, unless it is compelled to change that state by forces impressed thereon.

Newton's second law: A nett force F accelerates a body in the direction of the force. To any body may be ascribed a scalar constant, its mass m, such that the acceleration a produced in two bodies by a given force is inversely proportional to their masses, ie, for a given F, $a_2/a_1 = m_1/m_2$.

The alteration of motion is ever proportional to the motive force impressed; and is made in the direction of the right line in which that force is impressed.

Newton's third law: Forces always come in pairs, and the sum of the pair is zero.

To every action there is always opposed an equal and opposite reaction: or the mutual actions of two bodies upon each other are always equal, and directed to contrary parts.

Newton's Laws as Equations

When we look at forces specifically (as we do later for electricity and magnetism), we always define forces in pairs. There is an electrical attraction between an electron and a proton, and the force on the proton due to the electron equals that on the proton due to the electron but is in the opposite direction. The gravitational force the earth exerts on the moon equals that exerted by the moon on the earth. Or in general, for bodies 1 and 2, and for any interaction between them,

Newton's third law: $F_{1,2} = -F_{2,1}$.

If we write the total force acting a body as F, its mass as m and its acceleration as a, then the we may write

Newton's first and second law combined: $F = ma$.

For a body of finite size, one might ask what acceleration is meant. In this case, a is the acceleration of its centre of mass.

NEWTON'S THIRD LAW AND THE ANTHROPIC PRINCIPLE

Usually, scientific laws are developed by induction from experiment. They cannot be proved. For that reason, the work-energy theorem (the kinetic energy $mv^2/2$ of a point mass is increased by an amount equal to the total work done on it) is not a law: it is a theorem, a logical consequence of Newton's second law and some simple calculus. However, there is an indirect 'explanation', if not a proof, of Newton's third law.

Consider an object (make the example concrete: say a pen). According to Newton's third law, the force exerted by the bottom half of the pen on the top half plus that exerted by the top on the bottom add to zero. Now imagine that the third law is false, and that these two forces together produce a non-zero force on the pen. So it accelerates, and continues to do so. A universe in which objects spontaneously accelerated to large relative velocities would be a very dangerous place to live, mammals would not have evolved, and one of them would not be sitting here writing obscure footnotes. This is an example of the weak anthropic principle: if you find that the universe has laws and/or conditions that allow the existence of a being such as yourself, then you should not be surprised.

LAWS OF MOTION

Newton's Laws of Motion are still used by physicists all over the world. Newtonian Physics is taught in many courses at the college level, as well as high school and middle school. Everything in that genre of physics is based upon these three laws:

1. Every object has uniform motion unless acted upon by a force.
2. The force applied to an object is equal to the object's mass times the resulting acceleration: $F = m.a$
3. For every action, there is an equal and opposite reaction.

These laws are used to describe everything from throwing a ball to the merging of galaxies. Even though it has been shown through people like Albert Einstein that Newton's Laws are really approximations, they work so well on a wide range of scales, that they are still used in lieu of the better laws. For example, you would use Newtonian Mechanics to describe the acceleration of a car, not Einstein's General Relativity.

LAW OF GRAVITY

Newton's Law of Gravity is not precise in extreme circumstances, such as very high velocities or very strong gravity. For cases such as these, Einstein's General and Special Relativity theories are needed. However, in most other cases, and especially those that we are familiar with on Earth, Newton's Law works extremely well.

It is based upon his laws of motion, and it shows how two objects exhibit a force upon the other. It is the equation to the right.

It says that the gravitational force experienced is equal to a gravitational constant times both masses divided by the distance between them squared. The value "*G*" is an extremely small number, and therefore the gravitational force is extremely weak - the weakest of the four fundamental forces. This law also shows that the force of gravity dies off with the square of the distance. This means that if you are twice as far away from something, then the gravitational force you experience is 1/4 as much. The relationship between force and motion was expressed by Sir Isaac Newton in his three laws of motion: (1) a body at rest tends to remain at rest or a body in motion tends to remain in motion at a constant speed in a straight line unless acted on by an outside force, i.e., if the net unbalanced force is zero, then the acceleration is zero; (2) the acceleration a of a mass m by an unbalanced force F is directly proportional to the force and inversely proportional to the mass, or $a = F/m$; (3) for every action there is an equal and opposite reaction. The third law implies that the total momentum of a system of bodies not acted on by an external force remains constant (Refer conservation laws, in physics for more details). Newton's laws of motion, together with his law of gravitation, provide a satisfactory basis for the explanation of motion of everyday macroscopic objects under everyday conditions. However, when applied to extremely high speeds or extremely small objects, Newton's laws break down. Motion at speeds approaching the speed of light must be described by the theory of relativity. The equations derived from the theory of relativity reduce to Newton's when the speed of the object being described is very small compared to that of light. When the motions of extremely small objects (atoms and elementary particles) are described, the wavelike properties of matter must be taken into account (refer quantum theory). The theory of relativity also resolves the question of absolute motion. When one speaks of an object as being in motion, such motion is usually in reference to another object which is considered at rest. Although a person sitting in a car is at rest with respect to the car, both in motion with respect to the earth, and the earth is in motion with respect to the sun and the centre of the galaxy. All these motions are relative. It was once thought that there existed a light-carrying medium, known as the luminiferous ether, which was in a state of absolute rest. Any object in motion with respect to this hypothetical frame of reference would be in absolute motion. The theory of relativity showed, however, that no such medium was necessary and that all motion could be treated as relative.

NEWTON VS EINSTEIN

Einstein extended his Special Relativity theory to include gravitation and non-uniform motion. Einstein was intrigued by the fact that the two ways of measuring mass come up with the same value. In Newton's second law of motion, an object's mass is measured by seeing how much it resists a change in motion (its inertia). In Newton's law of gravity, an object's mass is determined by measuring how much gravity force it feels. The fact that the two masses are the same is why Galileo found that all things will fall with the same acceleration.

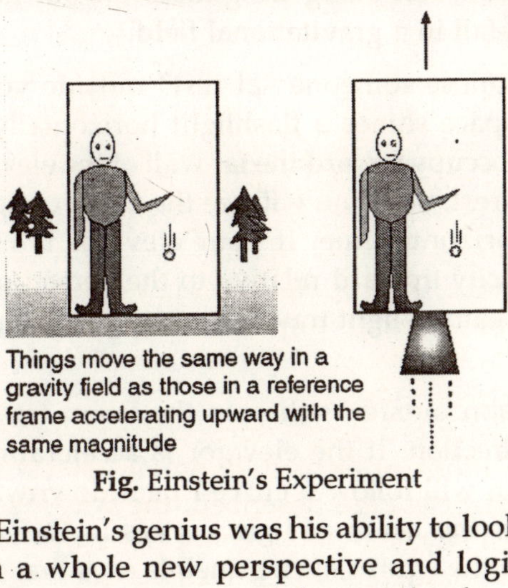

Fig. Einstein's Experiment

Part of Einstein's genius was his ability to look at ordinary things from a whole new perspective and logically follow through on the consequence of the insights he gained from his new perspective. He proposed an experiment involving two elevators: one at rest on the ground on the Earth and another, far out in space away from any planet, moon, or star, accelerating upward with an acceleration equal to that of one Earth gravity (9.8 metres/second^2). (Modern readers can substitute "rocket ship" for Einstein's elevator.) If a ball is dropped in the elevator at rest on the Earth, it will accelerate toward the floor with an acceleration of 9.8 metres/second^2. A ball released in the upward accelerating elevator far out in space will also accelerate toward the floor at 9.8 metres/second^2. The two elevator experiments get the same result!

Einstein used this to formulate the equivalence principle that would be the foundation of General Relativity. It states that "there is no experiment a person could conduct in a small volume of space that would distinguish between a gravitational field and an equivalent uniform acceleration". A consequence of this is that if an elevator is falling freely toward the ground because of gravity, an occupant inside will feel weightless just as if the elevator was far away from any planet, moon, or star. No experiment would help you distinguish between being weightless far out in space and being in free-fall in a gravitational field.

Now suppose someone "at rest" outside your elevator way out in space shines a flashlight horizontally across the elevator you occupy toward the far wall of the elevator. If your elevator is at rest, then you will see the beam of light travel in a straight horizontal line. If your elevator is moving at a constant velocity upward relative to the person outside, you will see the beam of light travel in a straight-line path angled downward.

The person outside still sees the beam travelling in a horizontal direction. If the elevator is accelerating upward, then the beam will follow a curved path downward relative to you. But if the beam of light curves in the accelerating elevator, then the equivalence principle says that the beam of light should also follow a curved path in a gravitational field.

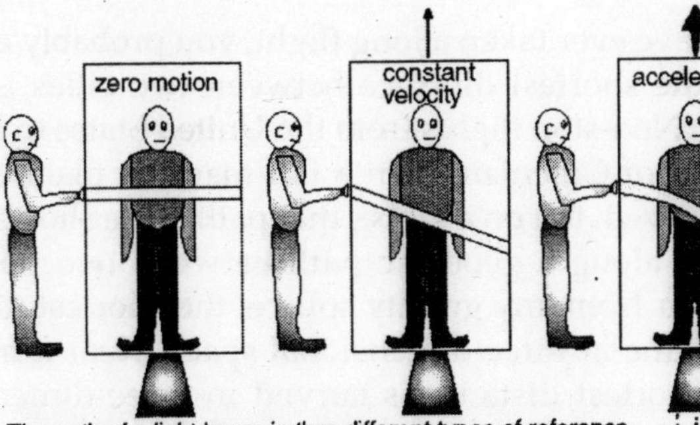

The path of a light beam in thre different types of reference frames moving with respect to the person *outside* the elevator. The light path shown is what the person *inside* the elevator sees. Under large acceleration, the beam of light will curve downward. It should also do that in a region of strong gravity.

Fig: Einstien Experiment

Light travels along the shortest path between two points in spacetime (a *geodesic).* If the geodesic is curved, then the path of light is curved. Einstein proposed in his General Relativity theory that what is called gravity is really the result of curved spacetime.

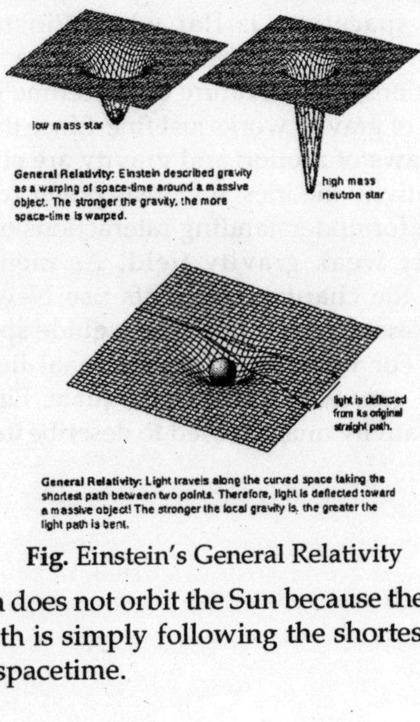

Fig. Einstein's General Relativity

The Earth does not orbit the Sun because the Sun is pulling on it. The Earth is simply following the shortest path in four-dimensional spacetime.

If you have ever taken a long flight, you probably already know that the shortest distance between two cities is not a straight line. Non-stop flights from the United States to Europe fly over parts of Greenland. On a flat map the plane's flight path looks curved, but on a globe, that path is the shortest one! Light travels along a geodesic path between two points in spacetime. Far from any gravity source, the shortest distance is a straight line in three-dimensional space. Near a massive object, the shortest distance is curved in three-dimensional space. Stephen Hawking gives the nice analogy that what we see is like the curved motion of a shadow on the ground from a plane flying in a straight line over hilly terrain.

Einstein's theory of General Relativity is a continuation or extension of Newton's law of gravity. Einstein's theory is not perfect (no scientific theory is absolutely perfect), but it does give a better understanding of the universe. In weak gravity conditions, they will give essentially the same results or predictions. Newton's law of gravity assumes that the geometry of spacetime is flat while Einstein's General Relativity allows any geometry to apply to spacetime. In weak gravity conditions, the curvature of spacetime is so small that Newton's law of gravity works just fine. Since the mathematics of Newton's laws of motion and gravity are simpler than for Einstein's relativity theories, scientists prefer to use Newton's law of gravity for understanding interactions of slow-moving objects in any weak gravity field. As mentioned at the beginning of the chapter, scientists use Newton's laws of motion and gravity to very accurately guide spacecraft in our solar system. For very strong gravitational fields, Newton's description of gravity becomes inadequate. Einstein's theory of General Relativity must be used to describe the gravitational effects.

Chapter 2

Basic Principles

FROM EXPERIENCE TO SPACETIME

"I might revel in the world of intelligibility which still remains to me, but although I have an idea of this world, yet I have not the least knowledge of it, nor can I ever attain to such knowledge with all the efforts of my natural faculty of reason. It is only a something that remains when I have eliminated everything belonging to the senses... but this something I know no further... There must here be a total absence of motive - unless this idea of an intelligible world is itself the motive... but to make this intelligible is precisely the problem that we cannot solve".

Immanuel Kant

We ordinarily take for granted the existence through time of objects moving according to fixed laws in three-dimensional space, but this is a highly abstract model of the objective world, far removed from the raw sense impressions that comprise our actual experience. This model may be consistent with our sense impressions, but it certainly is not uniquely determined by them. For example, Ptolemy and Copernicus constructed two very different conceptual models of the heavens based on essentially the same set of raw sense impressions. Likewise Weber and Maxwell synthesized two very different conceptual models of electromagnetism to account for a single set of observed phenomena. The fact that our raw sense impressions and experiences are (at least nominally) compatible with widely differing concepts of the world has led some philosophers to suggest that we should dispense with the idea of an "objective world" altogether, and base our physical theories on nothing but direct sense impressions, all else being merely the products of our imaginations. Berkeley expressed the positivist identification of sense impressions with objective existence by the famous phrase "esse est percipi" (to be is to be perceived). However, all attempts to base physical theories on nothing but raw sense impressions, avoiding arbitrary conceptual elements, invariably founder at the very start, because we have no sure means of distinguishing sense impressions from our thoughts and ideas. In fact, even the decision to make such a distinction represents a significant conceptual choice, one that is not strictly necessary on the basis of experience.

The process by which we, as individuals, learn to recognize sense impressions induced by an external world, and to distinguish them from our own internal thoughts and ideas, is highly complicated, and perhaps ultimately inexplicable. As Einstein put it (paraphrasing Kant) "the eternal mystery of the world is its comprehensibility". Nevertheless, in order to examine the epistemological foundations of any physical theory, we must give some consideration to how the elements of the theory are actually derived from our raw sense impressions, without automatically interpreting them in conventional terms. On the other hand, if we suppress every pre-conceived notion, including ordinary rules of reasoning, we can hardly hope to make any progress. We must choose a level of abstraction deep enough to give a meaningful perspective, but not so deep that it can never be connected to conventional ideas.

As an example of a moderately abstract model of experience, we might represent an idealized observer as a linearly ordered sequence of states, each of which is a function of the preceding states and of a set of raw sense impressions from external sources. This already entails two profound choices. First, it is a purely passive model, in the sense that it does not invoke volition or free will. As a result, all conditional statements in this model must be interpreted only as correlations, because without freedom it is meaningless to talk about the different consequences of alternate hypothetical actions. Second, by stipulating that the states are functions of the preceding but not the subsequent states we introduce an inherent directional asymmetry to experience, even though the justification for this is far from clear.

Still another choice must be made as to whether the sequence of states and experiences is continuous or discrete. In either case we can parameterize the sequence by a variable λ, and for the sake of definiteness we might represent each state $S(\lambda)$ and the corresponding sense impressions $E(\lambda)$ by strings of binary bits. Now, because of the mysterious comprehensibility of the world, it may happen that some functions of S are correlated with some functions of E. (Since this is a passive model by assumption, we cannot assert anything more than statistical correlations, because we do not have the freedom to arbitrarily vary S and determine the resulting E, but in principle we could still passively encounter enough variety of states and experiences to infer the most prominent correlations). These most primitive correlations are presumably "hard-wired" into higher-level categories of senses and concepts (i.e., state variables), rather than being sorted out cognitively. In terms of these higher-level variables we might find that over some range of l the sense impressions $E(\lambda)$ are strictly correlated with three functions θ, ϕ, α of the state $S(\lambda)$, which change only incrementally from one state to the next. Also, we may find that E is only incrementally different for incremental differences in θ, ϕ, α (independent of the prior values of those functions), and that this is the smallest and simplest set of functions with this property. Finally, suppose the sense impressions corresponding to a given set of values of the state functions are identical if the values of those functions are increased or decreased by some constant.

This describes roughly how an abstract observer might infer an orientation space along with the associated modes of interaction. In conventional terms, the observer infers the existence of external objects which induce a particular set of sense impressions depending on the observer's orientation. (Of course, this interpretation is necessarily conjectural; there may be other, perhaps more complex, interpretations that correspond as well or better with the observer's actual sequence of experiences.) At some point the observer may begin to perceive deviations from the simple three-variable orientation model, and find it necessary to adopt a more complicated conceptual model in order to accommodate the sequence of sense impressions. It remains true that the simple orientation model applies over sufficiently small ranges of states, but the sense impressions corresponding to each orientation may vary as a function of three additional state variables, which in conventional terms represent the spatial position of the observer. Like the orientation variables, these translation variables, which we might label x, y, and z, change only incrementally from one state to the next, but unlike the orientation variables there is no apparent periodicity.

Note that the success of this process of induction relies on a stratification of experiences, allowing the orientation effects to be discerned first, more or less independent of the translation effects. Then, once the orientation model has been established, the relatively small deviations from it (over small ranges of the state variable) could be interpreted as the effects of translatory motion. If not for this stratification (either in magnitude or in some other attribute), it might never be possible to infer the distinct sources of variation in our sense impressions. (On a more subtle level, the detailed metrical aspects of these translation variables will also be found to differ from those of the orientation variables, but only after quantitative units of measure and coordinates have been established.)

Another stage in the development of our hypothetical observer might be prompted by the detection of still more complicated variations in the experiential attributes of successive states. The observer may notice that while most of the orientation space is consistent with a fixed position, some particular features of their sense impressions do not maintain their expected relations to the other features, and no combination of the observer's translation and orientation variables can restore consistency. The inferred external objects of perception can no longer be modeled based on the premise that their relations with respect to each other are unchanging. Significantly, the observer may notice that some features vary as would be expected if the observer's own positional state had changed in one way, whereas other features vary as would be expected if the observer's positions had changed in a different way. From this recognition the observer concludes that, just as he himself can translate through the space, so also can individual external objects, and the relations are reciprocal. Thus, to each object we now assign an independent set of translation coordinates for each state of the observer.

In so doing we have made another important conceptual choice, namely, to regard "external objects" as having individual identities that persist from one state to the next. Other interpretations are possible. For example, we could account for the apparent motion of objects by supposing that one external entity simply ceases to exist, and another similar entity in a slightly different position comes into existence. According to this view, there would be no such thing as motion, but simply a sequence of arrangements of objects with some similarities. This may seem obtuse, but according to quantum mechanics it actually is not possible to unambiguously map the identities of individual elementary particles (such as electrons) from one event to another (because their wave functions overlap). Thus the seemingly innocuous assumption of continuous and persistent identities for material objects through time is actually, on some level, demonstrably false. However, on the macroscopic level, physical objects do seem to maintain individual identities, or at least it is possible to successfully model our sense impressions based on the assumption of persistent identities (because the overlaps between wave functions are negligible), and this success is the justification for introducing the concept of motion for the objects of experience.

The conceptual model of our hypothetical observer now involves something that we may call distance, related to the translational state variables, but it's worth noting that we have no direct perception of distances between ourselves and the assumed external objects, and even less between one external object and another. We have only our immediate sense impressions, which are understood to be purely local interactions, involving signals of some kind impinging on our senses. We infer from these signals a conceptual model of space and time within which external objects reside and move. This model actually entails two distinct kinds of extent, which we may call distance and length. An object, consisting of a locus of sense impressions that maintains a degree of coherence over time, has a spatial length, as do the paths that objects may follow in their motions, but the conceptual model of space also allows us to conceive of a distance between two objects, defined as the length of the shortest possible path between them.

The task of quantifying these distances, and of relating the orientation variables with the translation variables, then involves further assumptions. Since this is a passive model, all changes are strictly known only as a function of the single state variable, but we imagine other pseudo-independent variables based on the observed correlations. We have two means of quantifying spatial distances. One is by observing the near coincidence of one or more stable entities (measuring rods) with the interval to be quantified, and the other is to observe the change in the internal state variable as an object of stable speed moves from one end of the interval to the other. Thus we can quantify a spatial interval in terms of some reference spatial interval, or in terms of the associated temporal interval based on some reference state of motion. We identify these references purely by induction based on experience.

Combining the rotational symmetries and the apparent translational distances that we infer from our primary sense impressions, we conventionally arrive at a conception of the external world that is, in some sense, the dual of our subjective experience. In other words, we interpret our subjective experience as a one-dimensional temporally-ordered sequence of events, whereas we conceive of "the objective world now" corresponding to a single perceived event as a three-dimensional expanse of space as illustrated below:

In this way we intuitively conceive of time and space as inherently perpendicular dimensions, but complications arise if we posit that each event along our subjective path resides in, and is an element of, an objective world. If the events along any path are discrete, then we might imagine a simple sequence of discrete "instantaneous worlds":

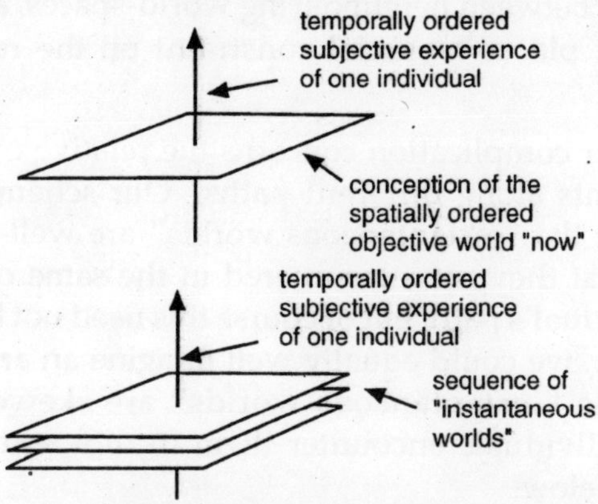

Fig. Instantaneous Worls

One difficulty with this arrangement is that it isn't clear how (or whether) these worlds interact with each other. If we regard each "instant" as a complete copy of the spatial universe, separate from every other instant, then there seems to be no definite way to identify an object in one world with "the same" object in another, particularly considering qualitatively identical objects such as electrons. If we have two electrons assigned the labels A and B in one instant of time, and if we find two electrons in the next instant of time, we have no certain way of deciding which of them was the "A" electron from the previous instant. (In fact, we cannot even map the spatial locations of one instant to "the same" locations in any other instant.) This illustrates how the classical concept of motion is necessarily based on the assumption of persistent identities of objects from one instant to another. Since it does seem possible (at least in the classical realm) to organize our experiences in terms of individual objects with persistent and unambiguous identities over time, we may be led to suspect that the sequence of existence of an individual or object in any one instant must be, in some sense, connected to or contiguous with its existence in neighbouring instants. If these objects are the constituents of "the world", this suggests that space itself at any "instant" is continuous with the spaces of neighbouring instants. This is important because it implies a definite connectivity between neighbouring world-spaces, and this, as we will see, places a crucial constraint on the relativity of motion.

Another complication concerns the relative orderings of world-instants along different paths. Our schematic above implied that the "instantaneous worlds" are well-ordered in the sense that they are encountered in the same order along every individual's path, but of course this need not be the case. For example, we could equally well imagine an arrangement in which the "instantaneous worlds" are skewed, so that different individuals encounter them in different orders, as illustrated below.

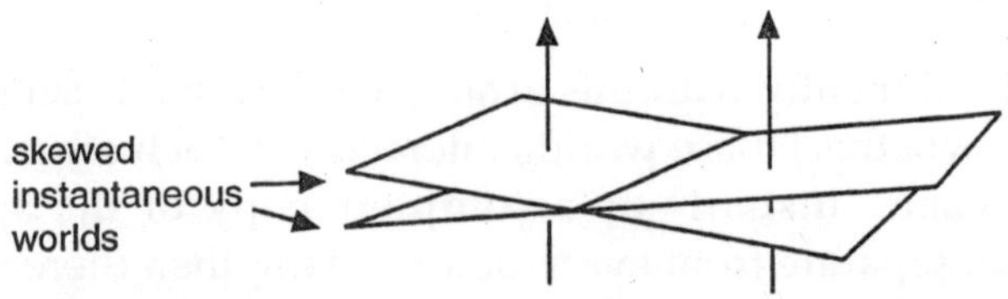

The concept of motion assumes the world can be analyzed in two different ways, first as the union of a set of mutually exclusive "events", and second as a set of "objects" each of which participates in an ordered sequence of events. In addition to this ordering of events encountered by each individual object, we must also assume both a co-lateral ordering of the events associated with different objects, and a transverse ordering of events from one object to another. These three kinds of orderings are illustrated schematically below.

This diagram suggests that the idea of motion is actually quite complex, even in this simple abstract model. Intuitively we regard motion as something like the derivative of the spatial "position" with respect to "time", but we can't even unambiguously define the distance between two worldlines, because it depends on how we correlate the temporal ordering along one line to the temporal ordering along the other. Essentially our concept of motion is overly ambitious, because we want it to express the spatial distance from the observer to the object for each event along the observer's worldline, but the intervals from one worldline to another are not confined to the worldlines themselves, so we have no definite way of assigning those intervals to events along our worldline. The best we can do is correlate all the intervals from a particular point on the observer's worldline to the object's worldline.

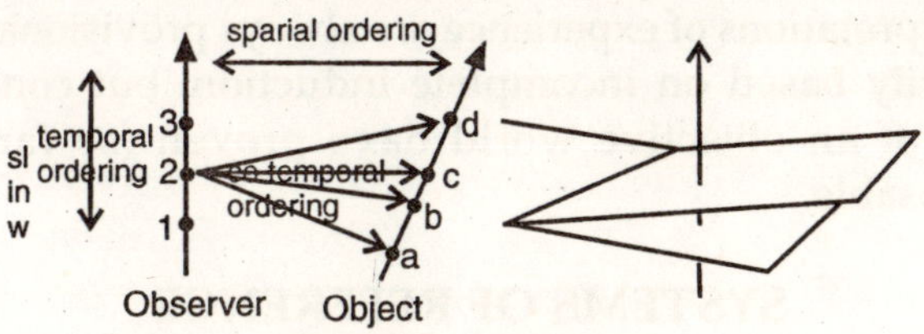

Fig. Idea of Motion

When we considered everything in terms of the sense impressions of just a single observer this was not an issue, since only one parameterization was needed to map the experiences of that observer, interpreted solipsistically. Any convenient parameterization was suitable. When we go on to consider multiple observers and objects we can still allow each observer to map his experiences and internal states using the most convenient terms of reference (which will presumably include his own state-index as the temporal coordinate), but now the question arises as to how all these private coordinate systems are related to each other. To answer this question we need to formalize our parameterizations into abstract systems of coordinates, and then consider how the coordinates of any given event with respect to one system are related to the coordinates of the same event with respect to another system. This is discussed in the coming section.

Considering how far removed from our raw sense impressions is our conceptual model of the external world, and how many unjustified assumptions and interpolations are involved in its construction, it's easy to see why some philosophers have advocated the rejection of all conceptual models. However, the fact remains that the imperative to reconcile our experience with some model of an objective external world has been one of the most important factors guiding the development of physical theories. Even in quantum mechanics, arguably the field of physics most resistant to complete realistic reconciliation, we still rely on the "correspondence principle", according to which the observables of the theory must conform to the observables of classical realistic models in the appropriate limits. Naturally our interpretations of experience are always provisional, being necessarily based on incomplete induction, but conceptual models of an objective world have proven (so far) to be indispensable.

SYSTEMS OF REFERENCE

If there were no solid bodies in nature there would be no geometry.

Henri Poincare

There are many theories of relativity, each of which can be associated with some arbitrariness in our descriptions of events. For example, suppose we describe the spatial relations between stationary particles on a line by assigning a real-valued coordinate to each particle, such that the distance between any two particles equals the difference between their coordinates. There is a degree of arbitrariness in this description due to the fact that all the coordinates could be increased by some arbitrary constant without affecting any of the relations between the particles. Symbolically this translational relativity can be expressed by saying that if x is a suitable system of coordinates for describing the relations between the particles, then so is x + k for any constant k. Likewise if we describe the spatial relations between stationary particles on a plane by assigning an ordered pair of real-valued coordinates to each particle, such that the squared distance between any two particles equals the sum of the squares of the differences between their respective coordinates, then there is a degree of arbitrariness in the description (in addition to the translational relativity of each individual coordinate) due to the fact that we could rotate the coordinates of every particle by an arbitrary constant angle without affecting any of the relations between the particles. This relativity of orientation is expressed symbolically by saying that if (x,y) is a suitable system of coordinates for describing the positions of particles on a plane, then so is (ax-by, bx+ay) where $a^2 + b^2 = 1$.

These relativities are purely formal, in the sense that they are tautological consequences of the premises, regardless of whether they have any physical applicability. Our first premise was that it's possible to assign a single real-valued coordinate to each particle on a line such that the distance between any two particles equals the difference between their coordinates. If this premise is satisfied, the invariance of relations under coordinate transformations from x to x + k follows trivially, but if the pairwise distances between three given particles were, say, 5, 3, and 12 units, then no three numbers could be assigned to the particles such that the pairwise differences equal the distances. This shows that the n(n-1)/2 pairwise distances between n particles cannot be independent of each other if those distances can be encoded unambiguously by just n coordinates in one dimension or, more generally, by kn coordinates in k dimensions. A suitable system of coordinates in one dimension exists only if the distances between particles satisfy a very restrictive condition. Letting d(A,B) denote the signed distance from A to B, the condition that must be satisfied is that for every three particles A,B,C we have d(A,B) + d(B,C) + d(C,A) = 0. Of course, this is essentially the definition of co-linearity, but we have no a priori reason to expect this definition to have any applicability in the world of physical objects. The fact that it has wide applicability is a non-trivial aspect of our experience, albeit one that we ordinarily take for granted.

Likewise for particles in a region of three dimensional space the premise that we can assign three numbers to each particle such that the squared distance between any two particles equals the sum of the squares of the differences between their respective coordinates is true only under a very restrictive condition, because there are only 3n degrees of freedom in the n(n-1)/2 pairwise distances between n particles.

Just as we found relativity of orientation for the pair of spatial coordinates x and y, we also find the same relativity for each of the pairs x,z and y,z in three dimensional space. Thus we have translational relativity for each of the four coordinates x,y,z,t, and we have rotational relativity for each pair of spatial coordinates (x,y), (x,z), and (y,z). This leaves the pairs of coordinates (x,t), (y,t) and (z,t). Not surprisingly we find that there is an analogous arbitrariness in these coordinate pairs, which can be expressed (for the x,t pair) by saying that the relations between the instances of particles on a line as a function of time are unaffected if we replace the x and t coordinates with ax – bt and –bx + at respectively, where $a^2 - b^2 = 1$. These transformations (rotations in the x,t plane through an imaginary angle), which characterize the theory of special relativity, are based on the premise that it is possible to assign pairs of values, x and t, to each instance of each particle on the x axis such that the squared spacetime distance equals the difference between the squares of the differences between the respective coordinates.

Each of the above examples represents an invariance of physically measurable relations under certain classes of linear transformations. Extending this idea, Einstein's general theory of relativity shows how the laws of physics, suitably formulated, are invariant under an even larger class of transformations of space and time coordinates, including non-linear transformations, and how these transformations subsume the phenomena of gravity. In general relativity the metrical properties of space and time are not constant, so the simple premises on which we based the primitive relativities described above turn out not to be satisfied globally. However, it remains true that those simple premises are satisfied locally, i.e., over sufficiently small regions of space and time, so they continue to be of fundamental importance.

As mentioned previously, the relativities described above are purely formal and tautological, but it turns out that each of them is closely related to a non-trivial physical symmetry. There exists a large class of identifiable objects whose lengths maintain a fixed proportion to each other under the very same set of transformations that characterize the relativities of the coordinates. In other words, just as we can translate the coordinates on the x axis without affecting the length of any object, we also find a large class of objects that can be individually translated along the x axis without affecting their lengths. The same applies to rotations and boosts. Such changes are physically distinct from purely formal shifts of the entire coordinate system, because when we move individual objects we are actually changing the relations between objects, since we are moving only a subset of all the coordinated objects. (Also, moving an object from one stationary position to another requires acceleration.) Thus for each formal arbitrariness in the system of coordinates there exists a physical symmetry, i.e., a large class of entities whose extents remain in constant proportions to each other when subjected individually to the same transformations.

We refer to these relations as physical symmetries rather than physical invariances, because (for example) we have no basis for asserting that the length of a solid object or the duration of a physical process is invariant under changes in position, orientation or state of motion. We have no way of assessing the truth of such a statement, because our measures of length and duration are all comparative. We can say only that the spatial and temporal extents of all the "stable" physical entities and processes are affected (if at all) in exactly the same proportion by changes in position, orientation, and state of motion. Of course, given this empirical fact, it is often convenient to speak as if the spatial and temporal extents are invariant, but we shouldn't forget that, from an epistemological standpoint, we can assert only symmetry, not invariance.

In his original presentation of special relativity in 1905 Einstein took measuring rods and clocks as primitive elements, even though he realized the weakness of this approach. He later wrote of the special theory:

> It is striking that the theory introduces two kinds of physical things, i.e., (1) measuring rods and clocks, and (2) all other things, e.g., the electromagnetic field, the material point, etc. This, in a certain sense, is inconsistent; strictly speaking, measuring rods and clocks should emerge as solutions of the basic equations (objects consisting of moving atomic configurations), not, as it were, as theoretically self-sufficient entities. The procedure was justified, however, because it was clear from the very beginning that the postulates of the theory are not strong enough to deduce from them equations for physical events sufficiently complete and sufficiently free from arbitrariness to form the basis of a theory of measuring rods and clocks.

This is quite similar to the view he expressed many years earlier

> ...the solid body and the clock do not in the conceptual edifice of physics play the part of irreducible elements, but that of composite structures, which may not play any independent part in theoretical physics. But it is my conviction that in the present stage of development of theoretical physics these ideas must still be employed as independent ideas; for we are still far from possessing such certain

knowledge of theoretical principles as to be able to give exact theoretical constructions of solid bodies and clocks.

The first quote is from his Autobiographical Notes in 1949, whereas the second is from his essay on Geometry and Experience published in 1921. It's interesting how little his views had changed during the intervening 28 years, despite the fact that those years saw the advent of quantum mechanics, which many would say provided the very theoretical principles underlying the construction of solid bodies and clocks that Einstein felt had been lacking. Whether or not the principles of quantum mechanics are adequate to justify our conceptions of reference lengths and time intervals, the characteristic spatial and temporal extents of quantum phenomena are used today as the basis for all such references.

Considering the arbitrariness of absolute coordinates, one might think our spatio-temporal descriptions could be better expressed in purely relational terms, such as by specifying only the mutual distances (minimum path lengths) between objects. Nevertheless, the most common method of description is to assign absolute coordinates (three spatial and one temporal) to each object, with reference to an established system of coordinates, while recognizing that the choice of coordinate systems is to some extent arbitrary. The relations between objects are then inferred from these absolute (thought somewhat arbitrary) coordinates. This may seem to be a roundabout process, but there are several reasons for using absolute coordinate systems to encode the relations between objects, rather than explicitly specifying the relations themselves.

One reason is that this approach enables us to take advantage of the efficiency made possible by the finite dimensionality of space. As discussed earlier if there were no limit to the dimensionality of space, then we would expect a set of n particles to have n(n-1)/2 independent pairwise spatial relations, so to explicitly specify all the distances between particles would require n-1 numbers for each particle, representing the distances to each of the other particles. Fcr a large number of particles (to say nothing of a potentially infinite number) this would be impractical. Fortunately the spatial relations between the objects of our experience are not mutually independent. The nth particle essentially adds only three (rather than n-1) degrees of freedom to the relational configuration. In physical terms this restriction can be clearly seen from the fact that the maximum number of mutually equidistant particles in D-dimensional space is D+1. Experience teaches us that in our physical space we can arrange four, but not five or more, particles such that they are all mutually equidistant, so we conclude that our space has three dimensions.

Historically the use of absolute coordinates rather than explicit relations may also have been partly due to the fact that analytic geometry and Cartesian coordinates were invented (by Fermat, Descartes and others) at almost the same time that the new science of mechanics needed them, just as tensor analysis was invented, three hundred years later, at the very moment when it was needed to facilitate the development of general relativity. (Of course, such coincidences are not accidental; contrivances requiring new materials tend to be invented soon after the material becomes available.) The coordinate systems of Descartes were not merely efficient, they were also consistent with the ancient Aristotelian belief (also held by Descartes) that there is no such thing as empty space or vacuum, and that continuous substance permeates the universe. In this context we cannot even contemplate explicitly specifying each individual distance between substantial points, because space is regarded as a continuum of substance. For Aristotle and Descartes, every spatial extent is a measure of the length of some substance, not a pure distance between particles as contemplated by atomists. In this sense we can say that the continuous absolute coordinate systems inherited by modern science from Aristotle and Descartes are a remnant of the Cartesian natural philosophy.

Another, perhaps more compelling, reason for the adoption of abstract coordinate systems in the descriptions of physical phenomena was the need to account for acceleration. As Newton explained with the example of a "spinning pail", the mutual relations between a set of material particles in an instant are not adequate to fully characterize a physical situation – at least not if we are considering only a small subset of all the particles in the universe. (Whether the mutual relations would be adequate if all the matter in the universe was taken into account is an open question.) In retrospect, there were other possible alternatives, such as characterizing not just the relations between particles at a specific instant, but over some temporal span of existence, but this would have required the unification of spatial and temporal measures, which did not occur until much later. Originally the motions of objects were represented simply by allowing the spatial coordinates of each persistent object to be continuous single-valued functions of one real variable, the time coordinate.

Incidentally, one consequence of the use of absolute coordinates is that it automatically entails a breaking of the alleged translational symmetry. We said previously that the coordinate system x could be replaced by $x + k$ for any real number k, implying that every real value of k is in some sense equally suitable. However, from a strictly mathematical point of view there does not exist a uniform distribution over the real numbers, so this form of representation does not exactly entail the perfect symmetry of position in an infinite space, even if the space is completely empty.

The set of all combinations of values for the three spatial coordinates and one time coordinate is assumed to give a complete coordination not only of the spatial positions of each entity at each time, but of all possible spatial positions at all possible times. Any definite set of space and time coordinates constitutes a system of reference. There are infinitely many distinct ways in which such coordinates can be assigned, but they are not entirely arbitrary, because we limit the range of possibilities by requiring contiguous physical entities to be assigned contiguous coordinates. This imposes a definite structure on the system, so it is more than merely a set of labels; it represents the most primitive laws of physics.

One way of specifying an entire model of a world consisting of n (classical) particles would be to explicitly give the 3n functions $x_j(t)$, $y_j(t)$, $z_j(t)$ for j = 1 to n. In this form, the un-occupied points of space would be irrelevant, since only the actual paths of actual physical entities have any meaning. In fact, it could be argued that only the intersections of these particles have physical significance, so the paths followed by the particles in between their mutual intersections could be regarded as merely hypothetical. Following this approach we might end up with a purely combinatorial specification of discrete interactions, with no need for the notion of a continuous physical space within which entities reside and move. However, the hypothesis that physical objects have continuous positions as functions of time with respect to a specified system of reference has proven to be extremely useful, especially for purposes of describing simple laws by which the observable interactions can be efficiently described and predicted.

An important class of physical laws that make use of the full spatio-temporal framework consists of laws that are expressed in terms of fields. A field is regarded as existing at each point within the system of coordinates, even those points that are not occupied by a material particle. Therefore, each continuous field existing throughout time has, potentially, far more degrees of freedom than does a discrete particle, or even infinitely many discrete particles. Arguably, we never actually observe fields, were merely observe effects attributed to fields. It's ironic that we can simplify the descriptions of particles by introducing hypothetical entities (fields) with far more degrees of freedom, but the laws governing the behaviour of these fields (e.g., Maxwell's equations for the electromagnetic field) along with symmetries and simple boundary conditions suffice to constrain the fields so that actually do provide a simplification. (Fields also provide a way of maintaining conservation laws for interactions "at a distance".) Whether the usefulness of the concepts of continuous space, time, and fields suggests that they possess some ontological status is debatable, but the concepts are undeniably useful.

These systems of reference are more than simple labeling. The numerical values of the coordinates are intended to connote physical properties of order and measure. In fact, we might even suppose that the sequence of states of all particles are uniformly parameterized by the time coordinate of our system of reference, but therein lies an ambiguity, because it isn't clear how the temporal states of one particle are to be placed in correspondence with the temporal states of another. Here we must make an important decision about how our model of the world is to be constructed. We might choose to regard the totality of all entities as comprising a single element in a succession of universal temporal states, in which case the temporal correspondence between entities is unambiguous. In such a universe the temporal coordinate induces a total ordering of events, which is to say, if we let the symbol £ denote temporal precedence or equality, then for every three events a,b,c we have

i. $a \le a$

ii. if $a \le b$ and $b \le a$, then $a = b$

iii. if $a \le b$ and $b \le c$, then $a \le c$

iv. either $a \le b$ or $b \le a$

However, this is not the only possible choice. We might choose instead to regard the temporal state of each individual particle as an independent quantity, bearing in mind that orderings of the elements of a set are not necessarily total. For example, consider the subsets of a flat plane, and the ordering induced by the inclusion relation $\subseteq$. Obviously the first three axioms of a total ordering are satisfied, because for any three subsets a,b,c of the plane we have (i) $a \subseteq a$, (ii) if $a \subseteq b$ and $b \subseteq a$, then $a = b$, and (iii) if $a \subseteq b$ and $b \subseteq c$, then $a \subseteq c$. However, the fourth axiom is not satisfied, because it's entirely possible to have two sets neither of which is included in the other. An ordering of this type is called a partial ordering, and we should allow for the possibility that the temporal relations between events induce a partial rather than a total ordering. In fact, we have no a priori reason to expect that temporal relations induce even a partial ordering. It is safest to assume that each entity possesses its own temporal state, and let our observations teach us how those states are mutually related, if at all. (Similar caution should be applied when modeling the relations between the spatial states of particles.)

Given any system of space and time coordinates we can define infinitely many others such that speeds are preserved. This represents an equivalence relation, and we can then define a reference frame as an equivalence class of coordinate systems such that the speed of each object has the same value in terms of each coordinate system in that class. Thus within a reference frame we can speak of the speed of an object, without needing to specify any particular coordinate system. Of course, just as our coordinate systems are generally valid only locally, so too are the reference frames.

Purely kinematic relativity contains enough degrees of freedom that we can simply define our systems of reference (i.e., coordinate systems) to satisfy the additivity of velocity. In other words, we can adopt velocity additivity as a principle, and this is essentially what scientists had tacitly done since ancient times. The great insight of Galileo and his successors was that this principle is inadequate to single out the physically meaningful reference systems.

INERTIA AND RELATIVITY

These or none must serve for reasons, and it is my great happiness that examples prove not rules, for to confirm this opinion, the world yields not one example.

John Donne

In his treatise "On the Revolution of Heavenly Spheres" Copernicus argued for the conceivability of a moving Earth by noting that

."..every apparent change in place occurs on account of the movement either of the thing seen or of the spectator, or on account of the necessarily unequal movement of both. No movement is perceptible relatively to things moved equally in the same direction - I mean relatively to the thing seen and the spectator".

This is a purely kinematical conception of relativity, like that of Aristarchus, based on the idea that we judge the positions (and changes in position) of objects only in relation to the positions of other objects. Many of Copernicus's contemporaries rejected the idea of a moving Earth, because we do not directly "sense" any such motion. To answer this objection, Galileo developed the concept of inertia, which he illustrated by "thought experiment" involving the behaviour of objects inside a ship which is moving at some constant speed in a straight line. He pointed out that

... "among things which all share equally in any motion, [that motion] does not act, and is as if it did not exist... in throwing something to your friend, you need throw it no more strongly in one direction than in another, the distances being equal... jumping with your feet together, you pass equal spaces in every direction."..

Thus Galileo's approach was based on a dynamical rather than a merely kinematic analysis, because he refers to forces acting on bodies, asserting that the dynamic behaviour of bodies is homogeneous and isotropic in terms of (suitably defined) measures in any uniform state of motion. This soon led to the modern principle of inertial relativity, although Galileo himself seems never to have fully grasped the distinction between accelerated and unaccelerated motion. He believed, for example, that circular motion was a natural state that would persist unless acted upon by some external agent. This shows that the resolution of dynamical behaviour into inertial and non-inertial components - which we generally take for granted today - is more subtle than it may appear. As Newton wrote:

"..the whole burden of philosophy seems to consist in this: from the phenomena of motions to infer the forces of nature, and then from these forces to deduce other phenomena..."

Newton's doctrine implicitly assumes that forces can be inferred from the motions of objects, but establishing the correspondence between forces and motions is not trivial, because the doctrine is, in a sense, circular. We infer "the forces of nature" from observed motions, and then we account for observed motions in terms of those forces. This assumes we can distinguish between forced and unforced motion, but there is no a priori way of making such a distinction. For example, the roughly circular motion of the Moon around the Earth might suggest the existence of a force (universal gravitation) acting between these two bodies, but it could also be taken as an indication that circular motion is a natural form of unforced motion, as Galileo believed. Different definitions of unforced motion lead to different sets of implied "forces of nature". The task is to choose a definition of unforced motion that leads to the identification of a set of physical forces that gives the most intelligible decomposition of phenomena. By indirect reasoning, the natural philosophers of the seventeenth century eventually arrived at the idea that, in the complete absence of external forces, an object would move uniformly in a straight line, and that, therefore, whenever we observe an object whose speed or direction of motion is changing, we can infer that an external force - proportional to the rate of change of motion - is acting upon that object. This is the principle of inertia, the most successful principle ever proposed for organizing our knowledge of the natural world. Notice that it refers to how a free object "would" move, because no object is completely free from all external forces. Thus the conditions of this fundamental principle, as stated, are never actually met, which highlights the subtlety of Newton's doctrine, and the aptness of his assertion that it comprises "the whole burden of philosophy". Also, notice that the principle of inertia does not discriminate between different states of uniform motion in straight lines, so it automatically entails a principle of relativity of dynamics, and in fact the two are essentially synonymous.

The first explicit statement of the modern principle of inertial relativity was apparently made by Pierre Gassendi, who is most often remembered today for reviving the ancient Greek doctrine of atomism. In the 1630's Gassendi repeated many of Galileo's experiments with motion, and interpreted them from a more abstract point of view, consciously separating out gravity as an external influence, and recognizing that the remaining "natural states of motions" were characterized not only by uniform speeds (as Galileo had said) but also by rectilinear paths. In order to conceive of inertial motion, it is necessary to review the whole range of observable motions of material objects and imagine those motions if the effects of all known external influences were removed. From this resulting set of ideal states of motion, it is necessary to identify the largest possible "equivalence class" of relatively uniform and rectilinear motions. These motions and configurations then constitute the basis for inertial measurements of space and time, i.e., inertial coordinate systems. Naturally inertial motions will then necessarily be uniform and rectilinear with respect to these coordinate systems, by definition.

Shortly thereafter (1644), Descartes presented the concept of inertial motion in his "Principles of Philosophy":

> Each thing...continues always in the same state, and that which is once moved always continues to move...and never changes unless caused by an external agent... all motion is of itself in a straight line...every part of a body, left to itself, continues to move, never in a curved line, but only along a straight line.

Similarly, in Huygens' "The Motion of Colliding Bodies" (composed in the mid 1650's but not published until 1703), the first hypothesis was that Any body already in motion will continue to move perpetually with the same speed in a straight line unless it is impeded.

Ultimately Newton incorporated this principle into his masterpiece, "Philosophiae Naturalis Principia Mathematica" (The Mathematical Principles of Natural Philosophy), as the first of his three "laws of motion"

1. Every body continues in its state of rest, or of uniform motion in a right line, unless it is compelled to change that state by the forces impressed upon it.

2. The change of motion is proportional to the motive force impressed, and is made in the direction of the right line in which that force is impressed.
3. To every action there is always opposed an equal and opposite reaction; or, the mutual actions of two bodies upon each other are always equal, and directed to contrary parts.

These "laws" expresses the classical mechanical principle of relativity, asserting equivalence between the conditions of "rest" and "uniform motion in a right line". Since no distinction is made between the various possible directions of uniform motion, the principle also implies the equivalence of uniform motion in all directions in space. Thus, if everything in the universe is a "body" in the sense of this law, and if we stipulate rules of force (such as Newton's second and third laws) that likewise do not distinguish between bodies at rest and bodies in uniform motion, then we arrive at a complete system of dynamics in which, as Newton said, "absolute rest cannot be determined from the positions of bodies in our regions". Corollary 5 of the Newton's Principia states:

> The motions of bodies included in a given space are the same among themselves, whether that space is at rest or moves uniformly forwards in a straight line without circular motion.

Of course, this presupposes that the words "uniformly" and "straight" have unambiguous meanings. Our concepts of uniform speed and straight paths are ultimately derived from observations of inertial motions, so the "laws of motion" are to some extent circular. These laws were historically expressed in terms of inertial coördinate systems, which are defined as the coordinate systems in terms of which these laws are valid. In other words, we define an inertial coordinate system as a system of space and time coordinates in terms of which inertia is homogeneous and isotropic, and then we announce the "laws of motion", which consist of the assertion that inertia is homogeneous and isotropic with respect to inertial coordinate systems. Thus the "laws of motion" are true by definition.

Their significance lies not in their truth, which is trivial, but in their applicability. The empirical fact that there exist systems of inertial coordinates is what makes the concept significant. We have no a priori reason to expect that such coordinate systems exist, i.e., that the forces of nature would resolve themselves so coherently on this (or any other finite) basis, but they evidently do. In fact, it appears that not just one such coordinate system exists (which would be remarkable enough), but that infinitely many of them exist, in all possible states of relative motion. To be precise, the principle of relativity asserts that for any material particle in any state of motion there exists an inertial coordinate system in terms of which the particle is (at least momentarily) at rest.

It's important to recognize that Newton's first law, by itself, is not sufficient to identify the systems of coordinates in terms of which all three laws of motion are satisfied. The first law serves to determine the shape of the coordinate axes and inertial paths, but it does not fully define a system of inertial coordinates, because the first law is satisfied in infinitely many systems of coordinates that are not inertial. The system of oblique xt coordinates illustrated below is an example of such a system.

The two dashed lines indicate the paths of two identical objects, both initially at rest with respect to these coordinates and propelled outward from the origin by impulses forces of equal magnitude (acting against each other). Every object not subject to external forces moves with uniform speed in a straight line with respect to this coordinate system, so Newton's First Law of motion is satisfied, but the second law clearly is not, because the speeds imparted to these identical objects by equal forces are not equal. In other words, inertia is not isotropic with respect to these coordinates. In order for Newton's Second Law to be satisfied, we not only need the coordinate axes to be straight and uniformly graduated relative to freely moving objects, we need the space axes to be aligned in time such that mechanical inertia is the same in all spatial directions (so that, for example, the objects whose paths are represented by the two dashed lines in the above figure have the same speeds). This effectively establishes the planes of simultaneity of inertial coordinate systems. In an operational sense, Newton's Third Law is also involved in establishing the planes of simultaneity for an inertial coordinate system, because it is only by means of the Third Law that we can actually define "equal forces" as the forces necessary to impart equal "quantities of motion" (to use Newton's phrase). Of course, this doesn't imply that inertial coordinate systems are the "true" systems of reference. They are simply the most intuitive, convenient, and readily accessible systems, based on the inertial behaviour of material objects.

In addition to contributing to the definition of an inertial coordinate system, the third law also serves to establish a fundamental aspect of the relationships between relatively moving inertial coordinate systems. Specifically, the third law implies (requires) that if the spatial origin of one inertial coordinate system is moving at velocity v with respect to a second inertial coordinate system, then the spatial origin of the second system is moving at velocity -v with respect to the first. This property is sometimes called reciprocity, and is important for the various derivations of the Lorentz transformation to be presented in subsequent sections.

Based on the definition of an inertial coordinate system, and the isotropy of inertia with respect to such coordinates, it follows that two identical objects, initially at rest with respect to those coordinates and exerting a mutual forc e on each other, recoil by equal distances in equal times (in accord with Newton's third law). Assuming the lengths of stable material objects are independent of their spatial positions and orientations (spatial homogeneity and isotropy), it follows that we can synchronize distant clocks with identical particles ejected with equal forces from the mid-point between the clocks. Of course, this operational definition of simultaneity is not new. It is precisely what Galileo described in his illustration of inertial motion onboard a moving ship. When he wrote that an object thrown with equal force will reach equal distances [in the same time], he was implicitly defining simultaneity at separate locations on the basis of inertial isotropy. This is crucial to understanding the significance of inertial coordinate systems. The requirement for a particular object to be at rest with respect to the system suffices only to determine the direction of the "time axis", i.e., the loci of constant spatial position. Galileo and his successors realized (although they did not always explicitly state) that it is also necessary to specify the loci of constant temporal position, and this is achieved by choosing coordinates in such a way that mechanical inertia is isotropic. (This means the inertia of an object does not depend on any absolute reference direction in space, although it may depend on the velocity of the object. It is sufficient to say the resistance to acceleration of a resting object is the same in all spatial directions.)

Conceptually, to establish a complete system of space and time coordinates based on inertial isotropy, imagine that at each point in space there is an identically constructed cannon, and all these cannons are at rest with respect to each other. At one particular point, which we designate as the origin of our coordinates, is a clock and numerous identical cannons, each pointed at one of the other cannons out in space. The cannons are fired from the origin, and when a cannonball passes one of the external cannons it triggers that external cannon to fire a reply back to the origin. Each cannonball has identifying marks so we can correlate each reply with the shot that triggered it, and with the identity of the replying cannon. The ith reply event is assigned the time coordinate $t_i = [t_{return}(i) - t_{send}(i)]/2$ seconds, and it is assigned space coordinates x_i, y_i, z_i based on the angular direction of the sending cannon and the radial distance $r_i = t_i$ cannon-seconds. This procedure would have been perfectly intelligible to Newton, and he would have agreed that it yields an inertial coordinate system, suitable for the application of his three laws of motion.

Naturally given one such system of coordinates, we can construct infinitely many others by simple spatial re-orientation of the space axes and/or translation of the spatial or temporal axes. All such transformations leave the speed of every object unchanged. An equivalence class of all such inertial coordinate systems is called an inertial reference frame. For characterizing the mutual dynamical states of two material bodies, the associated inertial rest frames of the bodies are more meaningful than the mere distance between the bodies, because any inertial coordinate system possesses a fixed spatial orientation with respect to any other inertial coordinate system, enabling us to take account of tangential motion between bodies whose mutual distance is not changing. For this reason, the physically meaningful "relative velocity of two material bodies" is best defined as their reciprocal states of motion with respect to each others' associated inertial rest frame coordinates.

The principle of relativity does not tell us how two relatively moving systems of inertial coordinates are related to each other, but it does imply that this relationship can be determined empirically. We need only construct two relatively moving systems of inertial coordinates and compare them. Based on observations of coordinate systems with relatively low mutual speeds, and with the limited precision available at the time, Galileo and Newton surmised that if (x,t) is an inertial coordinate system then so is (x',t'), where

$$x' = x - vt \qquad\qquad t' = t$$

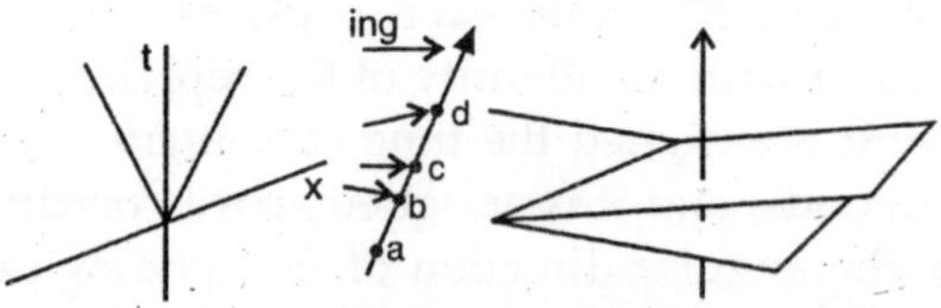

Fig. Inertia Equation

and v is the mutual speed between the origins of the two systems. This implies that relative speeds are simply additive. In other words, if a material object B is moving at the speed v in terms of inertial rest frame coordinates of A, and if an object C is moving in the same direction at the speed u in terms of inertial rest frame coordinates of B, then C is moving at the speed v + u in terms of inertial rest frame coordinates of A. This conclusion may seem plausible, but it's important to realize that we are not free to arbitrarily adopt this or any other transformation and speed composition rule for the set of inertial coordinate systems, because those systems are already fully defined (up to insignificant scale factors) by the requirements for inertia to be homogeneous and isotropic and for momentum to be conserved. These properties suffice to determine the set of inertial coordinate systems and (therefore) the relationships between them. Given these conditions, the relationship between relatively moving inertial coordinate systems, whatever it may be, is a matter of empirical fact.

Of course, inertial isotropy is not the only possible basis for constructing spacetime coordinate systems. We could impose a different constraint to determine the loci of constant temporal position, such as a total temporal ordering of events. However, if we do this, we will find that mechanical inertia is generally not isotropic in terms of the resulting coordinate systems, so the usual symmetrical laws of mechanics will not be valid in terms of those coordinate systems (at least not if restricted to ponderable matter). Indeed this was the case for the ether theories developed in the late 19th century, as discussed in subsequent sections. Such coordinate systems, while extremely awkward, would not be logically inconsistent. The choices we make to specify a coordinate system and to resolve spacetime intervals into separate spatial and temporal components are to some extent conventional, provided we are willing to disregard the manifest symmetry of physical phenomena. But since physics consists of identifying and understanding the symmetries of nature, the option of disregarding those symmetries does not appeal to most physicists.

By the end of the nineteenth century a new class of phenomena involving electric and magnetic fields had been incorporated into physics, and the concept of inertia was found to be applicable to these phenomena as well. For example, Maxwell's equations imply that a pulse of light conveys momentum. Hence the principle of inertia ought to apply to electromagnetism as well as to the motions of material bodies. In his 1905 paper "On the Electrodynamics of Moving Bodies" Einstein adopted this more comprehensive interpretation of inertia, basing the special theory of relativity on the proposition that:

> The laws by which the states of physical systems undergo changes are not affected, whether these changes of state be referred to the one or the other of two systems of [inertial] coordinates in uniform translatory motion.

This is nearly identical to Newton's Corollary 5. It's unfortunate that the word "inertial" was omitted, because, as noted above, uniform translatory motion is not sufficient to ensure that a system of coordinates is actually an inertial coordinate system. However, Einstein made it clear that he was indeed talking about inertial coordinate systems when he previously characterized them as coordinate systems "in which the equations of Newtonian mechanics hold good". Admittedly this is a somewhat awkward assertion in the context of Einstein's paper, because one of the main conclusions of the paper is that the equations of Newtonian mechanics do not precisely "hold good" with respect to inertial coordinate systems. Recognizing this inconsistency, Sommerfeld added a footnote in subsequent published editions of Einstein's paper, qualifying the statement about Newtonian mechanics holding good "to the first approximation", but this footnote does not really clarify the situation. Fundamentally, the class of coordinate systems that Einstein was trying to identify (the inertial coordinate systems) are those in terms of which inertia is homogeneous and isotropic, so that free objects move at constant speed in straight lines, and the force required to accelerate an object from rest to a given speed is the same in all directions. As discussed above, these conditions are just sufficient to determine a coordinate system in terms of which the symmetrical equations of mechanics hold good, but without pre-supposing the exact form of those equations.

Since light (i.e., an electromagnetic wave) carries momentum, and the procedure for constructing an inertial coordinate system described previously was based on the isotropy of momentum, it is reasonable to expect that pulses of light could be used in place of cannonballs, and we should arrive at essentially the same class of coordinate systems. In his 1905 paper this is how Einstein described the construction of inertial coordinate systems, implicitly asserting that the propagation of light is isotropic with respect to the same class of coordinate systems in terms of which mechanical inertia is isotropic. In this respect it might seem as if he was treating light as a stream of inertial particles, and indeed his paper on special relativity was written just after the paper in which he introduced the concept of photons. However, we know that light is not exactly like a stream of material particles, especially because we cannot conceive of light being at rest with respect to any system of inertial coordinates. The way in which light fits into the framework of inertial coordinate systems is considered in the next section. We will find that although the principle of relativity continues to apply, and the definition of inertial coordinate systems remains unchanged, the relationship between relatively moving systems of inertial coordinate systems must be different than what Galileo and Newton surmised.

THE RELATIVITY OF LIGHT

According to the theory of emission, the transmission of energy of light is effected by the actual transference of light-corpuscles... According to the theory of undulation, there is a material medium which fills the space between two bodies, and it is by the action of contiguous parts of this medium that the energy is passed on...

James Clerk Maxwell

Light is arguably the phenomenon of nature with which we have the most conscious experience, by means of our sense of vision, and yet throughout most of human history very little seems to have been known about how vision works.

Interestingly, from the very beginning there were at least two distinct concepts of light, existing side by side, as can be seen in some of the earliest known writings. For example, the description of creation in the biblical book of Genesis says light was created on the first day, and yet the sun, moon, and stars were not created until the fourth day "to give light upon the earth". Evidently the word "light" is being used to signify two different things on the first and fourth days. For another example, Plato argued in Timaeus that there are two kinds of "fire" involved in our sense of vision, one coming from inside ourselves, emanating as visual rays from our eyes to make contact with distant objects, and another, which he called "daylight", that (when present) surrounds the visual rays from our eyes and facilitates the conveyance of the visual images. These two kinds of "fire" correspond roughly with the later scholastic concepts of lux and lumen. The word lux was used to signify our visual sensations, whereas the word lumen referred to an external agent (such as light from the sun) that somehow participates in our sense of vision.

There was also, in ancient times, a competing theory of vision, according to which all objects naturally emit whole "images" (eidola) of themselves in small packets, and these enter our souls by way of our eyes. To account for our inability to see at night, it was thought that light from the sun or moon struck the objects and caused them to emit their images. This model of vision still entailed two distinct kinds of light: the facilitating illumination from the sun or moon, and the eidola emitted by ordinary objects. This somewhat awkward conception of vision was improved by Ibn al-Haitham and later by Kepler, who argued that it is not necessary to assume whole objects emit multiple copies of themselves; we can simply consider each tiny part of an object as the source of rays emanating in all directions, and a sub-set of these rays intersecting in the eye can be re-assembled into an image of the object.

Until the end of the 17th century there was no evidence to indicate that rays of light propagated at a finite speed, and they were often assumed to be instantaneous. Only in 1689 with Roemer's observations of the moons of Jupiter, and even more convincingly in 1728 with Bradley's discovery of stellar aberration, did it become clear that the rays of lumen propagate through space with a characteristic finite speed. This suggested that light, and the energy it conveys, must have some mode of existence during the interval of time between its emission and its absorption. Hence light became an entity or process in itself, rather than just a relation between entities, but again there were two competing notions as to the mode of existence. Two different analogies were conceived, based on the behaviour of ordinary material substances. Some thought light could be regarded as a stream of material corpuscles moving through empty space, whereas other believed light consists of undulations or waves in a pervasive material medium. Each of these analogies was consistent with some of the attributes of light, but neither could be reconciled fully with all the attributes. For example, if light consists of material corpuscles, then according to Galilean relativity there should be an inertial reference frame with respect to which light is at rest in a vacuum, whereas in fact we never observe light in a vacuum to be at rest, nor even noticeably slow, with respect to any inertial reference frame. On the other hand, if light is a wave propagating through a material medium, then the constituent parts of that medium should, according to Galilean relativity, behave inertially, and in particular should have a definite rest frame, whereas we find that light propagates best through regions (vacuum) in which there is no detectable material with a definite rest frame, and again we cannot conceive of light at rest in any inertial frame. Thus the behaviour of light defies realistic representation in terms of the behaviour of material substances within the framework of Galilean space and time, even if we consider just the classical attributes, let alone quantum phenomena.

By the end of the 19th century the inadequacy of both of the materialistic analogies for explaining the behaviour of light had become acute, because there was strong evidence that light exhibits two seemingly mutually exclusive properties. First,

Maxwell showed how light can be regarded as a propagating electromagnetic wave, and as such the speed of propagation is obviously independent of the speed of the source. Second, numerous experiments showed that light propagates at the same speed in all directions relative to the source, just as we would expect for streams of inertial corpuscles. Hence some of the attributes of light seemed to unequivocally support an emission theory, while others seemed just as unequivocally to support a wave theory. In retrospect it's clear that there was an underlying confusion regarding the terms of description, i.e., the systems of inertial coordinates, but this was far from clear at the time.

One of the first clues to unraveling the mystery was found in 1887, when Woldemar Voigt made a remarkable discovery concerning the ordinary wave equation. Recall that the wave equation for a time-dependent scalar field f(x,t) in one dimension is

$$\frac{\partial^2 \phi}{\partial x^2} = \frac{1}{u^2}\frac{\partial^2 \phi}{\partial t^2}$$

where u is the propagation speed of the wave. This equation was first studied by Jean d'Alembert in the 18th century, and it applies to a wide range of physical phenomena. In fact it seems to represent a fundamental aspect of the relationship between space, time, and motion, transcending any particular application. Traditionally it was considered to be valid only for a coordinate system x,t with respect to which the wave medium (presumed to be an inertial substance) is at rest and has isotropic properties, because if we apply a Galilean transformation to these coordinates, the wave equation is not satisfied with respect to the transformed coordinates. However, Galilean transformations are not the most general possible linear transformations. Voigt considered the question of whether there is any linear transformation that leaves the wave equation unchanged.

The general linear transformation between (X,T) and (x,t) is of the form x = AX + BT   t = CX + DT

for constants A,B,C,D. If we choose units of space and time so that the acoustic speed u equals 1, the wave equation in terms of (X,T) is simply $\partial^2\phi/\partial X^2 = \partial^2\phi/\partial T^2$. To express this equation in terms of the transformed (x,t) coordinates, recall that the total differential of f can be written in the form

$$d\phi = \frac{\partial \phi}{\partial x}dx + \frac{\partial \phi}{\partial t}dt$$

Also, at any constant T, the value of f is purely a function of X, so we can divide through the above equation by dX to give

$$\frac{\partial \phi}{\partial X} = \left(\frac{d\phi}{dX}\right)_T = \frac{\partial \phi}{\partial x}\left(\frac{dx}{dX}\right)_T + \frac{\partial \phi}{\partial t}\left(\frac{dt}{dX}\right)_T$$

$$= A\frac{\partial \phi}{\partial x} + C\frac{\partial \phi}{\partial t}$$

Taking the partial derivative of this with respect to X then gives

$$\frac{\partial^2 \phi}{\partial X^2} = A\frac{\partial^2 \phi}{\partial X \partial x} + C\frac{\partial^2 \phi}{\partial X \partial t}$$

Since partial differentiation is commutative, this can be written as

$$\frac{\partial^2 \phi}{\partial X^2} = A\frac{\partial}{\partial x}\left(\frac{\partial \phi}{\partial X}\right) + C\frac{\partial}{\partial t}\left(\frac{\partial \phi}{\partial X}\right)$$

Substituting the prior expression for $\partial\phi/dX$ and carrying out the partial differentiations gives an expression for $\partial^2\phi/\partial X^2$ in terms of partials of ϕ with respect to x and t. Likewise we can derive an expression for $\partial^2\phi/\partial T^2$. Substituting into the wave equation gives

$$A^2\frac{\partial^2 \phi}{\partial x^2} + 2AC\frac{\partial^2 \phi}{\partial x \partial t} + C^2\frac{\partial^2 \phi}{\partial t^2} = B^2\frac{\partial^2 \phi}{\partial x^2} + 2BD\frac{\partial^2 \phi}{\partial x \partial t} + D^2\frac{\partial^2 \phi}{\partial t^2}$$

This is equivalent to the condition that f(X,T) is a solution of the wave equation with respect to the X,T coordinates. Since the mixed partial generally varies along a path of constant second partial with respect to x or t, it follows that a necessary and sufficient condition for $\phi(x,t)$ to also be a solution of the wave equation in terms of the x,t coordinates is that the constants A,B,C,D of our linear transformation satisfy the relations

$$A^2 + C^2 = B^2 + D^2 \qquad AC = BD$$

Furthermore, the differential of the space transformation is dx = AdX + BdT, so an increment with dx = 0 satisfies dX/dT = -B/A. This represents the velocity at which the spatial origin of the x,t coordinates is moving relative to the X,T coordinates. We will refer to this velocity as v. We also have the inverse transformation from (X,T) to (x,t):

$$X = \frac{D}{AD-BC}x + \frac{-B}{AD-BC}t$$

$$T = \frac{-C}{AD-BC}x + \frac{A}{AD-BC}t$$

Proceeding as before, the differential of this space transformation gives dx/dt = B/D for the velocity of the spatial origin of the X,T coordinates with respect to the x,t coordinates, and this must equal -v. Therefore we have B = -Av = -Dv, and so A = D. It follows from the condition imposed by the wave equation that B = C, so both of these equal -Av. Our transformation can then be written in the form

$$x = A(X - vT) \qquad t = A(T - vX)$$

The same analysis shows that the perpendicular coordinates y and z of the transformed system must be given by

$$y = A\sqrt{1-v^2}\,Y \qquad z = A\sqrt{1-v^2}\,Z$$

In order to make the transformation formula for x agree with the Galilean transformation, Voigt chose A = 1, so he did not actually arrive at the Lorentz transformation, but nevertheless he had shown roughly how the wave equation could actually be relativistic – just like the dynamic behaviour of inertial particles - provided we are willing to consider a transformation of the space and time coordinates that differs from the Galilean transformation. Had he considered the inverse transformation

$$X = \frac{1}{A\left(1-v^2\right)}\left(x+vt\right) \qquad T = \frac{1}{A\left(1-v^2\right)}\left(t+vx\right)$$

he might have noticed that the determinant is $A^2(1-v^2)$, so to make this equal to 1 we must have $A = 1/(1-v^2)^{1/2}$, which not only implies y = Y and z = Z, but also makes the transformation formally identical to its inverse. In other words, he would have arrived at a completely relativistic framework for the wave equation. However, this was not Voigt's objective, and he evidently regarded the transformed coordinates x, y, z and t as merely a convenient parameterization for purposes of calculation, without attaching any greater significance to them.

Voigt's transformation was the first hint of how a wavelike phenomenon could be compatible with the principle of relativity, which is that there exist inertial coordinate systems in terms of which free motions are linear, inertia is isotropic, and every material object is instantaneously at rest with respect to one of these systems. None of this conflicts with the observed behaviour of light, because the motion of light is observed to be both linear and isotropic with respect to inertial coordinate systems. The fact that light is not at rest with respect to any system of inertial coordinates does not conflict with the principle of relativity if we agree that light is not a material object.

The incompatibility of light with the Galilean framework arises not from any conflict with the principle of relativity, but from the tacitly adopted empirical conclusion that two relatively moving systems of inertial coordinates are related to each other by Galilean transformations, so that the composition of co-linear speeds is simply additive. As discussed in the previous section, we aren't free to impose this assumption on the class of inertial coordinate systems, because they are fully determined by the requirement for inertia to be homogeneous and isotropic. There are no more adjustable parameters (aside from insignificant scale factors), so the composition of velocities with respect to relatively moving inertial coordinate systems is a matter to be determined empirically.

Recall from the previous section that, on the basis of slowly moving reference frames, Galileo and Newton had inferred that the composition of speeds was simply additive. In other words, if a material object B is moving at the speed v in terms of inertial rest frame coordinates of a material object A, and if an object C is moving in the same direction at the speed u in terms of inertial rest frame coordinates of B, then Newton found that object C has the speed v + u in terms of the inertial rest frame coordinates of A. Toward the end of the nineteenth century, more precise observations revealed that is not quite correct. It was found that the speed of object C in terms of inertial rest frame coordinates of A is not v + u, but rather $(v + u)/(1 + uv/c^2)$, where c is the speed of light in a vacuum.

Obviously these conclusions would be identical if the speed of light was infinitely great, which was still considered a real possibility in Galileo's day. Many people, including Descartes, regarded rays of light as instantaneous. Even Newton's Opticks, published in 1704, made allowances for the possibility that "light be propagated in an instant" (although Newton himself was persuaded by Roemer's observations that light has a finite speed). Hence it can be argued that the principles of Galileo and Einstein are essentially identical in both form and content. The only difference is that Galileo assessed the propagation of light to be "if not instantaneous then extraordinarily fast", and thus could neglect the term uv/c^2, especially since he restricted his considerations to the movements of material objects, whereas subsequently it became clear that the speed of light has a finite value, and it was necessary to take account of the uv/c^2 term when attempting to incorporating the motions of light and high-speed particles into the framework of mechanics.

The empirical correspondence between inertial isotropy and lightspeed isotropy can be illustrated by a simple experiment. Three objects, A, B, and C, at rest with respect to each other can be arranged so that one of them is at the midpoint between the other two (the midpoint having been determined using standard measuring rods at rest with respect to those objects). The two outer objects, A and C, are equipped with identical clocks, and the central object, B, is equipped with two identical cannons. Let the two cannons in the centre be fired simultaneously in opposite directions toward the two outer objects, and then at a subsequent time let object B emit a flash of light. If the arrivals of the cannonball and light coincide at A, then they also coincide at C, signifying that the propagation of light is isotropic with respect to the same system of coordinates in terms of which mechanical inertia is isotropic, as illustrated in the figure below.

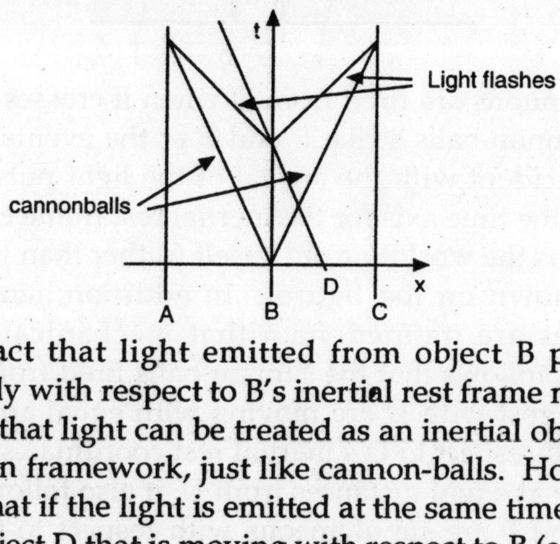

The fact that light emitted from object B propagates isotropically with respect to B's inertial rest frame might seem to suggest that light can be treated as an inertial object within the Galilean framework, just like cannon-balls. However, we also find that if the light is emitted at the same time and place from an object D that is moving with respect to B (as shown in the figure above), the light's speed is still isotropic with respect to B's inertial rest frame. Now, this might seem to suggest that light is a disturbance in a material medium in which the objects A,B,C just happen to be at rest, but this is ruled out by the fact that it applies regardless of the state of (uniform) motion of those objects. Naturally this implies that the flash of light propagates isotropically with respect to the inertial rest coordinates of object D as well. To demonstrate this, we could arrange for two other bodies, denoted by E and F, to be moving at the same speed as D, and located an equal distance from D in opposite directions. Then we could fire two identically constructed cannons (at rest with respect to D) in opposite directions, toward E and F. The results are illustrated below.

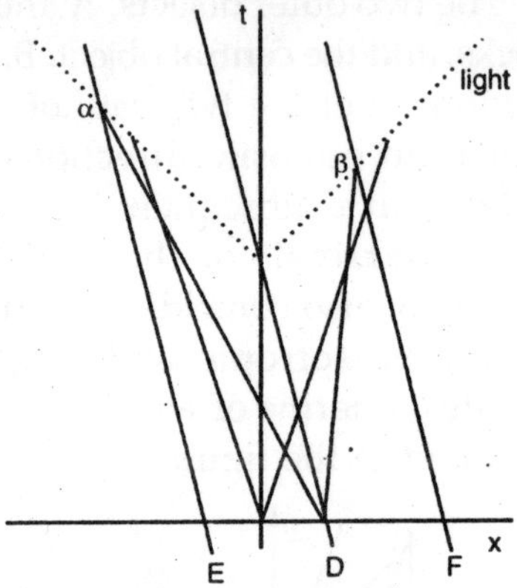

The cannons are fired from D when it crosses the x axis, and the cannon-balls strike E and F at the events marked α and β, coincident with the arrival of the light pulse from D. Obviously the time axis for the inertial rest frame coordinates of object D is the worldline of D itself (rather than the original "t" axis shown on the figure). In addition, since inertial coordinates are defined such that mechanical inertia is isotropic, it follows that the cannon-balls fired from identical cannons at rest with D are moving with equal and opposite speeds with respect to D's inertial rest coordinates, and since E and F are at equal distances from D, it also follows that the events a and b are simultaneous with respect to the inertial rest coordinates of D. Hence, not only is the time axis of D's rest frame slanted with respect to B's time axis, the spatial axis of D's rest frame is equally slanted with respect to B's spatial axis.

Several other important conclusions can be deduced from this figure. For example, with respect to the original x,t coordinate system, the speeds of the cannon-balls from D are not given by simply adding (or subtracting) the speed of the cannon-balls with respect to D's rest frame to (or from) the speed of D with respect to the x,t coordinates. Since momentum is explicitly conserved, this implies that the inertia of a body increases with it's velocity (i.e., kinetic energy), as is discussed in more detail in previous section. We should also note that although the speed of light is isotropic with respect to any inertial spacetime coordinates, independent of the motion of the source, it is not correct to say that the light itself is isotropic. The relationship between the frequency (and energy) of the light with respect to the rest frame of the emitting body and the frequency (and energy) of the light with respect to the rest frame of the receiving body does depend on the relative velocity between those two massive bodies .

Incidentally, notice that we can rule out the possibility of object B and D dragging the light medium along with them, because they are moving through the same region of space at the same time, and they can't both be dragging the same medium in opposite directions. This is in contrast to the case of (for example) acoustic pressure waves in a material substance, because in that case a recognizable material substance determines the unique isotropic frame, whereas in the case of light we're unable to identify any definite material medium, so the medium has no definite rest frame.

The first person to discern the true relationship between relatively moving systems of inertial coordinate systems was Hendrik Antoon Lorentz. Not surprisingly, he arrived at this conception in a rather indirect and laborious way, and didn't immediately recognize that the class of coordinate systems he had discovered (and which he called "local coordinate" systems) were none other than Galileo's inertial coordinate systems. Incidentally, although Lorentz and Voigt knew and corresponded with each other, Lorentz apparently was not aware of Voigt's earlier work on coordinate transformations that leave the wave equation invariant, and so that work had no influence on Lorentz's search for coordinate systems in terms of which Maxwell's equations are invariant. Unlike

Voigt, Lorentz derived the transformation in two separate stages. He first developed the "local time" coordinate, and only years later came to the conclusion (after, but independently of, Fitzgerald) that a "contraction" of spatial length was also necessary in order to account for the absence of second-order effects in Michelson's experiment.

Lorentz began with the absolute ether frame coordinates t and x, in terms of which every event can be assigned a unique space-time position (t,x), and then he considered a system moving with the velocity v in the positive x direction. He applied the traditional Galilean transformation to assign a new set of coordinates to every event. Thus an event with ether-frame coordinates t,x is assigned the new coordinates x" = x - vt and t" = t. Then he tentatively proposed an additional transformation that must be applied to x",t" in order to give coordinates in terms of which Maxwell's equations apply in their standard form. Lorentz was not entirely clear about the physical significance of these "local" coordinates, but it turns out that all physical phenomena conform to the same isotropic laws of physics when described in terms of these coordinates. (Lorentz's notation made use of the parameter $\beta = 1/\gamma = 1/(1-v^2)^{1/2}$ and another constant which he later determines to be 1.) Taking units such that c = 1, his equations for the local coordinates x' and t' in terms of the Galilean coordinates which we are calling x" and t" are

$$x' = \frac{x''}{y} \qquad t' = yt'' - \frac{vx''}{y}$$

Recall that the traditional Galilean transformation is x" = x - vt and t" = t, so we can make these substitutions to give the complete transformation from the original ether rest frame coordinates x,t to the local coordinates moving with speed v

$$x' = \frac{x - vt}{y} \qquad t' = \frac{t - vx}{y}$$

These effective coordinates enabled Lorentz to explain how two relatively moving observers, each using his own local system of coordinates, both seem to remain at the centre of expanding spherical light waves originating at their point of intersection, as illustrated below

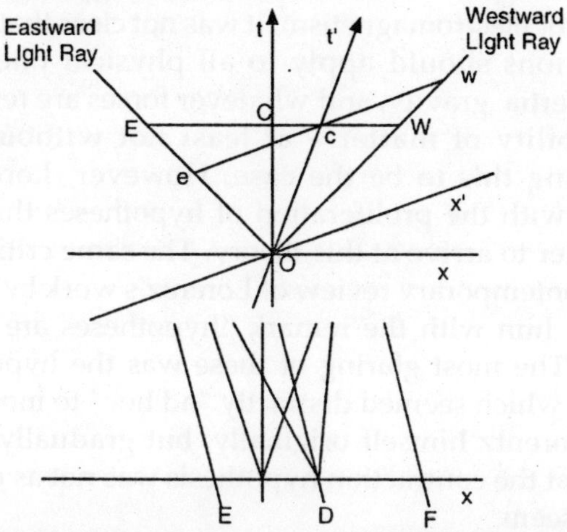

The x and x′ axes represent the respective spatial coordinates (say, in the east/west direction), and the t and t′ axes represent the respective time coordinates. One observer is moving through time along the t axis, and the other has some relative westward velocity as he moves through time along the t′ axis. The two observers intersect at the event labeled O, where they each emit a pulse of light. Those light pulses emanate away from O along the dotted lines. Subsequently the observer moving along the t axis finds himself at C, and according to his measures of space and time the outward going light waves are at E and W at that same instant, which places him at the midpoint between them. On the other hand, the observer moving along t′ axis finds himself at point c, and according to his measures of space and time the outward going light waves are at e and w at this instant, which implies that he is at the midpoint between them.

Thus Lorentz discovered that by means of the "fictitious" coordinates x′,t′ it was possible to conceive of a class of relatively moving coordinate systems with respect to which the speed of light is invariant. He went beyond Voigt in the realization that the existence of this class of coordinate systems ensures the appearance of relativity, at least for optical phenomena, and yet, like Voigt, he still tended to regard the "local coordinates" as artificial. Having been derived specifically for electromagnetism, it was not clear that the same transformations should apply to all physical phenomena, including inertia, gravity, and whatever forces are responsible for the stability of matter - at least not without simply hypothesizing this to be the case. However, Lorentz was dissatisfied with the proliferation of hypotheses that he had made in order to arrive at this theory. The same criticism was made in a contemporary review of Lorentz's work by Poincare, who chided him with the remark "hypotheses are what we lack least". The most glaring of these was the hypothesis of contraction, which seemed distinctly "ad hoc" to most people, including Lorentz himself originally, but gradually he came to realize that the contraction hypothesis was not as unnatural as it might seem.

Surprising as this hypothesis may appear at first sight, yet we shall have to admit that it is by no means far-fetched, as soon as we assume that molecular forces are also transmitted through the ether, like the electric and magnetic forces...

dly after the fact) that the Fitzgerald contraction was to be expected based on what he called the Molecular Force Hypothesis and his theorem of Corresponding States, as discussed in the next section.

Chapter 3

Complex of Relativity Phenomena

THE SPACETIME INTERVAL

The advance of the quantum wave function of any physical system as it passes uniformly from the event (t,x,y,z) to the event (t+dt, x+dx, y+dy, z+dz) is proportional to the value of dt given by

$$(d\tau)^2 = (dt)^2 - \frac{1}{c^2}\left[(dx)^2 + (dy)^2 + (dz)^2\right]$$

where t,x,y,z are any system of inertial coordinates and c is a constant (the speed of light, equal to 300 metres per microsecond). The quantity dt is called the elapsed proper time of the interval, and it is invariant with respect to any system of inertial coordinates. To illustrate, consider a muon particle, which has a radioactive mean life of roughly 2 msec with respect to its inertial rest frame coordinates. In other words, between the appearance of a typical muon (arising from, say, the decay of a pion) and its decay there is an interval of about 2 msec in terms of the time coordinate of the muon's inertial rest frame, so the components of this interval are {2,0,0,0}, and the quantum phase of the particle advances by an amount proportional to dt, where

$$d\tau = \sqrt{(2)^2 - \frac{1}{c^2}\left[(0)^2 - (0)^2 - (0)^2\right]} = 2\mu\text{sec}$$

Now suppose we assess this same physical phenomenon with respect to a relatively moving system of inertial coordinates, e.g., a system with respect to which the muon moved from the spatial origin [0,0,0] all the way to the spatial position [980m, -750m, 1270m] before it decayed. With respect to these coordinates, the muon traveled a spatial distance of 1771 metres. Since the advance of the quantum wave function (i.e., the proper time) of a system or particle over any interval of its worldline is invariant, the corresponding time component of this physical interval with respect to these relatively moving inertial coordinates must be much greater than 2 msec. If we let (dT,dX,dY,dZ) denote the components of this interval with respect to the relatively moving system of inertial coordinates, we must have

$$\sqrt{(dT)^2 - \frac{1}{c^2}\left[(dX)^2 + (dY)^2 + (dZ)^2\right]} = 2\mu\text{sec}$$

Solving for dT and substituting for the spatial components noted above, we have

$$dT = \sqrt{(2)^2 - \frac{1}{300^2}\left[(980)^2 + (-750)^2 + (1270)^2\right]} = 5.23\mu\text{sec}$$

This represents the time component of the muon decay interval with respect to the moving system of inertial coordinates. Since the muon has moved a spatial distance of 1771 metres in 6.23 msec, we see that its velocity with respect to these coordinates is 284 m/msec, which is 0.947c.

The identification of the spacetime interval with quantum phase applies to null intervals as well, consistent with the fact that the quantum phase of a photon does not advance at all between its emission and absorption. Hence the physical significance of a null spacetime interval is that the quantum state of any system is constant along that interval. In other words, the interval represents a single quantum state of the system. It follows that the emission and absorption of a photon must be regarded as, in some sense, a single quantum event.

Note, however, that the quantum phase is path dependent. In other words, two particles at opposite ends of a lightlike (null) interval do not share the same quantum state unless the second particle reached that event by passing along that null interval. Hence the concept of the spacetime interval as a measure of the phase of the quantum wave function does not conflict with the exclusion principle for fermions such as electrons, because even though two electrons can be null-separated, they cannot have separated along that null path, because they have non-zero rest mass. Of course, it is possible for two photons at opposite ends of a null interval to have reached that condition by progressing along that interval, in which case they represent the same quantum phase (and in some sense may be regarded as "the same photon"), but photons are bosons, and hence not excluded from occupying the same state. In fact, the presence of one photon in a particular quantum state actually enhances the probability of another photon entering that state. (This is responsible for the phenomenon of stimulated emission, which is the basis of operation of lasers.)

In this regard it's interesting to consider neutrinos, which (like electrons) are fermions, meaning that they have anti-symmetric eigenfunctions, and hence are subject to the Pauli exclusion principle. On the other hand, neutrinos were traditionally regarded as massless, meaning they propagate along null intervals. This raises the prospect of two instances of a neutrino at opposite ends of a null interval, with the second occupying the same quantum state as the first, in violation of the exclusion principle for fermions. It might be argued that these two instances are really the same neutrino, and a particle obviously can't exclude itself from occupying its own state. However, this is somewhat problematic due to the indistinguishability and the lack of definite identities for individual particles. A different approach would be to argue that all fermions, including neutrinos, must have mass, and thus be excluded from traveling along null intervals. The idea that neutrinos actually do have mass seems to be supported by recent experimental observations, but the questions remains open.

Based on the general identification of the invariant magnitude (proper time) of a timelike interval with quantum phase along that interval, it follows that all physical processes and characteristic sequences of events will evolve in proportion to this quantity. The name "proper time" is appropriate because this quantity represents the most meaningful known measure of elapsed time along that interval, based on the fact that the quantum state is the most complete possible description of physical reality. Since not all spacetime intervals are timelike, we conclude that the temporal relations between events induce only a partial ordering, rather than a total ordering, because a set of events can be totally ordered only if they are each inside the future or past null cone of each of the others. This doesn't hold if any of the pairwise intervals is spacelike. As a consequence of this partial ordering, between two fixed timelike separated events there exist timelike paths with different lapses of proper time.

Admittedly a partial ordering of events has been considered unacceptable by some people, basically because they regard total temporal ordering in a classical Cartesian setting as an inviolable first principle. Rather than accept partial ordering they prefer to (more or less arbitrarily) select one particular inertial reference system and declare it to be the "true" configuration, as in Lorentz's original theory, in an attempt to restore an unambiguous total temporal ordering to events. They then account for the apparent differences in elapsed time (as in muon observations) by regarding them as effects of absolute velocity relative to the "true" frame of reference, again following Lorentz. However, unlike Lorentz, we now have a theory of quantum mechanics, and the quantum state of a system gives (arguably) the most complete possible objective description of the system. Therefore, modern advocates of total temporal ordering face the daunting task of finding some mechanism underlying quantum mechanics (i.e., hidden variables) to provide a physical significance for their preferred total ordering. Unfortunately, the only prospects for a viable hidden-variable theory seem to be things like the explicitly non-local contrivances described by David Bohm, which must surely be anathema to those who seek a physics based on classical Cartesian mechanisms. So, although the theories of relativity and quantum mechanics are in some respects incongruent, it is nevertheless true that the (putative) validity and completeness of quantum mechanics constitutes one of the strongest argument in favour of the relativistic interpretation of Lorentz invariance.

We should also mention that a tacit assumption has been made above, namely, the assumption of physical equivalence between instantaneously co-moving frames, regardless of acceleration. For example, we assume that two co-moving clocks will keep time at the same instantaneous rate, even if one is accelerating and the other is not. This is just a hypothesis - we have no a priori reason to rule out physical effects of the 2nd, 3rd, 4th,... time derivatives. It just so happens that when we construct a theory on this basis, it works pretty well. (Similarly we have no a priori reason to think the field equations necessarily depend only on the metric and its 1st and 2nd derivatives; but it works.)

Another way of expressing this "clock hypothesis" is to say that an ideal clock is unaffected by acceleration, and to regard this as the definition of an "ideal clock", i.e., one that compensates for any effects of 2nd or higher derivatives. Of course the physical significance of this definition arises from the hypothesized fact that acceleration is absolute, and therefore perfectly detectable (in principle). In contrast, we hypothesize that velocity is perfectly undetectable, which explains why we cannot define our "ideal clock" to compensate for velocity (or, for that matter, position). The point is that these are both assumptions invoked by relativity: (1) the zeroth and first derivatives of position are perfectly relative and undetectable, and (2) the second and higher derivatives of position are perfectly absolute and detectable. Most treatments of relativity emphasize the first assumption, but the second is no less important.

The notion of an ideal clock takes on even more physical significance from the fact that there exist physical entities (such a vibrating atoms, etc) in which the intrinsic forces far exceed any accelerating forces we can apply, so that we have in fact (not just in principle) the ability to observe virtually ideal clocks. For example, in the Rebka and Pound experiments it was found that nuclear clocks were slowed by precisely the factor $\gamma(v)$, even though subject to accelerations up to 10^{16} g (which is huge in normal terms, but of course still small relative to nuclear forces).

It was emphasized in previous sections that a pulse of light has no inertial rest frame, but this may seem puzzling at first. The pulse has a well-defined spatial position versus time with respect to some inertial coordinate system, representing a fixed velocity c relative to that system, and we know that any system of orthogonal coordinates in uniform non-rotating motion relative to an inertial coordinate system is also inertial, so why can we not simply apply the velocity c to the base frame to arrive at the rest frame of the light pulse? How can an entity have a well-defined velocity and yet have no well-defined rest frame? The only answer can be that the transformation is singular, i.e., the coordinate system moving with a uniform speed c relative to an inertial frame is not well defined. The singular behaviour of the transformation corresponds to the fact that the absolute magnitude of the spacetime intervals along lightlike paths is null. The transformation through a velocity v from the xt to the x′t′ coordinates is $t' = (t-vx)/\gamma$ and $x' = (x-vt)/\gamma$ where $\gamma = (1-v^2)^{1/2}$, so it's clear that for $v = \pm 1$ the individual t′ and x′ components are undefined, but the ratio of dt′ over dx′ remains well-defined, with magnitude 1 and the opposite sign from v. The singularity of the Lorentz transformation for the speed c suggests that the conception of light as an entity in itself may be somewhat misleading, and it is often useful to regard light as simply an interaction between two massive bodies along a null spacetime interval.

Discussions of special relativity often refer to the use of clocks and reflected light signals for the evaluation of spacetime intervals. For example, suppose two identical clocks are moving uniformly with speeds +v and -v along the x axis of a given inertial coordinate system, and these clocks are set to zero at the intersection of their worldlines. When the leftward clock indicates the proper time t_1, it emits a pulse of light, which bounces off the rightward clock when that clock indicates τ_2, and arrives back at the leftward clock when that clock reads τ_3. This is illustrated in the drawing below.

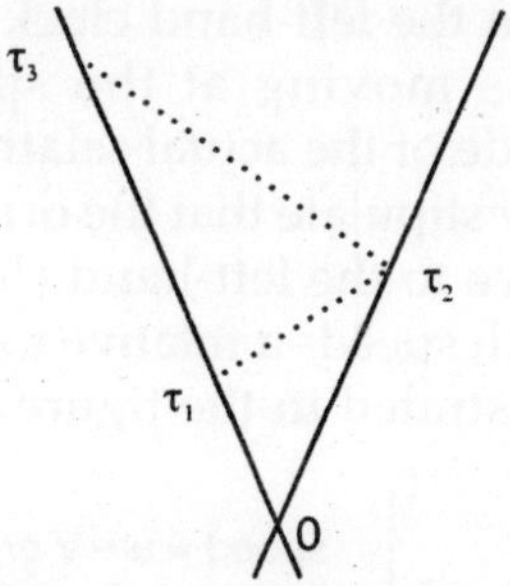

By similar triangles we immediately have $\tau_2/\tau_1 = \tau_3/\tau_2$, and thus $\tau_2^2 = \tau_1\tau_3$. Of course, this same relation holds good in Galilean spacetime as well (not to mention Euclidean plane geometry, using distances instead of time intervals), and the reflected signal need not be a light pulse. Any object moving at the same speed (angle) in both directions with respect to this coordinate system would serve just as well, and would lead to the same result that τ_2 is the geometric mean of τ_1 and τ_3. Naturally if we apply any Minkowskian, Galilean, or Euclidean transformation (respectively), the pictorial angles of the lines will differ, but the three absolute intervals will remain unchanged.

It is, of course, possible to distinguish between the Galilean and Minkowskian cases based just on the values of the elapsed times, provided we know the relative speeds of the clocks and the signal. In Galilean spacetime each proper time τ_j equals the coordinate time t_j, whereas in Minkowski spacetime it equals $(t_j^2 - x_j^2)^{1/2}$ where $x_j = \pm v\ t_j$. Hence the proper time τ_j in Minkowski spacetime is $t_j(1 - v^2)^{1/2}$. This might seem to imply that the ratios of proper times are the same in the Galilean and Minkowskian cases, but in fact we have not made a valid comparison for equal relative speeds between the clocks. In this example each clock is moving with speed v away from the midpoint, which implies that the relative speed is 2v in the Galilean case, but only $2v/(1 + v^2)$ in the Minkowskian case.

To give a valid comparison for equal relative speeds between the clocks, let's transform the events to a system of coordinate such that the left-hand clock is stationary and the right-hand clock is moving at the speed v. Now this v represents magnitude of the actual relative speed between the two clocks. We now stipulate that the original signal is moving with speed u relative to the left-hand clock, and the reflected signal is moving with speed -u relative to the right-hand clock. The situation is illustrated in the figure below.

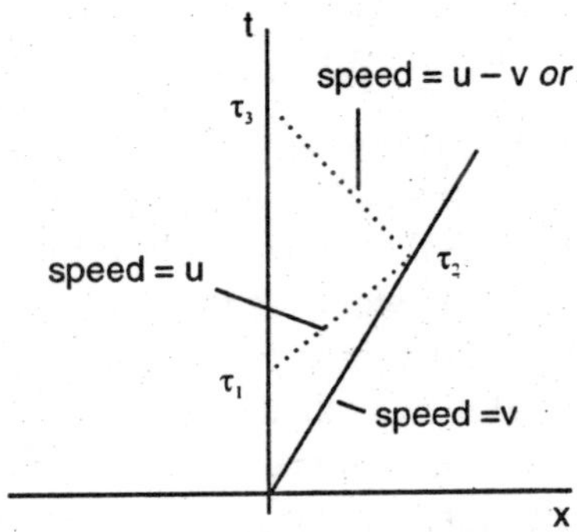

The speed, with respect to these coordinates, of the reflected signal is what distinguishes the Galilean from the Minkowskian case. Letting x_2 and t_2 denote the coordinates of the reflection event, and noting that $\tau_1 = t_1$ and $\tau_3 = t_3$, we have $v = x_2/t_2$ and $u = x_2/(t_2-\tau_1)$. We also have

$$\frac{x_2}{\tau_3 - t_2} = \begin{cases} u - v & \textit{Galilean} \\ (u-v)/(1-uv) & \textit{Minkowskian} \end{cases}$$

Dividing the numerator and denominator of the expression for u by t_2, and replacing x_2/t_2 with v, gives $u = v/[1-(t_1/t_2)]$. Likewise the above expressions can be written as

$$\frac{x_2}{(\tau_3 - t_2) - 1} = \begin{cases} u - v & \textit{Galilean} \\ (u-v)/(1-uv) & \textit{Minkowskian} \end{cases}$$

Solving these equations for the time ratios, we have

$$\frac{\tau_1}{t_2} = \frac{u-v}{u} \qquad \frac{\tau_3}{t_2} = \frac{u}{u-v} \quad or \quad \frac{u\left(1-v^2\right)}{u-v}$$

Consequently, depending on whether the metric is Galilean or Minkowskian, the ratio of t_3 over t_1 is given by

$$\frac{\tau_3}{\tau_1} = \left(\frac{u}{u-v}\right)^2 \quad \text{or} \quad \left(\frac{u}{u-v}\right)^2 \left(1-v^2\right)$$

respectively. If u happens to be unity (meaning that the signals propagate at the speed of light), these expressions reduce to the squares of the Galilean and relativistic Doppler shift factors, i.e., $1/(1\text{-}v)^2$ and $(1+v)/(1\text{-}v)$.

Another distinguishing factor between the two metrics is that with the Minkowski metric the speed of light is invariant with respect to any system of inertial coordinates, so (arguably) we can even say that it represents the same "u" relative to a spacelike interval as it does relative to a timelike interval, in order to adhere to our stipulation that the reflected signal has the speed u relative to "the rest frame of the right-hand clock". Of course, a spacelike interval cannot actually be the worldline of a clock (or any other material object), but the invariance of the speed of light under Minkowskian transformations enables us to rationally apply the same "geometric mean" formula to determine the magnitudes of spacelike intervals, provided we use light-like signals, as illustrated below.

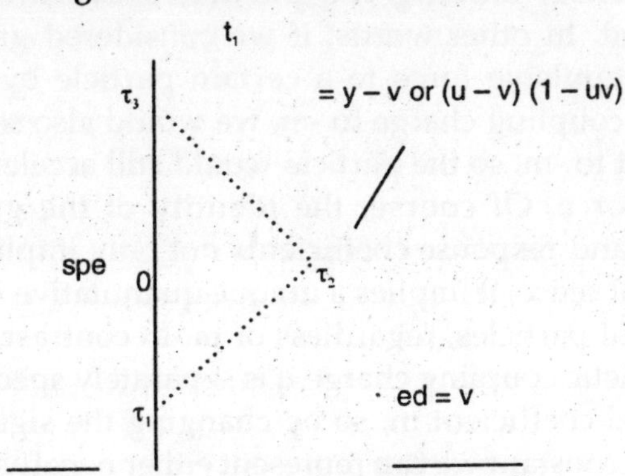

In this case we have $\tau_1 = -\tau_3$, so $\tau_2^2 = -\tau_3^2$, meaning that squared spacelike intervals are negative.

FORCE LAWS AND MAXWELL'S EQUATIONS

Perhaps the most rudimentary scientific observation is that material objects exhibit a natural tendency to move in certain circumstances. For example, objects near the surface of the Earth tend to move in the local "downward" direction, i.e., toward the Earth's centre. The Newtonian approach to describing such tendencies was to imagine a "force field" representing a vectorial force per unit charge that is applied to any particle at any given point, and then to postulate that the acceleration vector of each particle equals the applied force divided by the particle's inertial mass. Thus the "charge" of a particle determines how strongly that particle couples with a particular kind of force field, whereas the inertial mass determines how susceptible the particle's velocity is to arbitrary applied forces. In the case of gravity, the coupling charge happens to be the same as the inertial mass, denoted by m, but for electric and magnetic forces the coupling charge q differs from m.

Since the coupling charge and the response coefficient for gravity are identical, it follows that gravity can only operate in a single directional sense, because changing the sign of m for a particle would reverse the sense of both the coupling and the response, leaving the particle's overall behaviour unchanged. In other words, if we considered gravitation to apply a repulsive force to a certain particle by setting the particle's coupling charge to -m, we would also set its inertial coefficient to -m, so the particle would still accelerate into the applied force. Of course, the identity of the gravitational coupling and response coefficients not only implies a unique directional sense, it implies a unique quantitative response for all material particles, regardless of m. In contrast, the electric and magnetic coupling charge q is separately specifiable from the inertial coefficient m, so by changing the sign of q while leaving m constant we can represent either negative or positive response, and by changing the ratio of q/m we can scale the quantitative response.

According to this classical picture, a small test particle with mass m and electric charge q at a given location in space is subject to a vectorial force f given by

$$f = mg + qE + qv \times B \qquad \text{...(1a)}$$

where g is the gravitational field vector, E is the electric field vector, and B is the magnetic field vector at the given location, and v is the velocity vector of the test particle. (Refer previous sections for a review of vector products such as the cross product denoted by v ´ B.) As noted above, the acceleration vector a of the particle is simply f/m, so we have the equation of motion

$$a = g + \frac{q}{m}E + \frac{q}{m}v \times B \qquad \text{...(1b)}$$

Given the mass, charge, and initial position of a test particle, and the vectors g,E,B for every point in vicinity of the particle, this equation enables us to compute the particle's subsequent motion. Notice that acceleration of a test particle due to gravity is independent of the particle's properties and state of motion (to the first approximation), whereas the accelerations due to the electric and magnetic fields are both proportional to the particle's charge divided by it's inertial mass. In addition, the contribution of the magnetic field is a function of the particle's velocity. This dependence on the state of motion has important consequences, and leads naturally to the unification of the electric and magnetic fields, but before describing these effects it's worthwhile to briefly review the effect of the classical gravitational field on the motion of a particle.

The gravitational acceleration field g at a point p due to a distant particle of mass m was specified classically by Newton's law

$$g = -\frac{m}{r^3}r \qquad \text{...(2)}$$

Where r is the displacement vector (of magnitude r) from the mass particle to the point p. Noting that $r^2 = x^2 + y^2 + z^2$ and r = ix + jy + kz, it's straightforward to verify that the divergence of the gravitational field g vanishes at any point p away from the mass, i.e., we have

$$\nabla . g = o \qquad ...(3)$$

The field due to multiple mass particles is just the sum of the individual fields, so the divergence of g due to any configuration of matter vanishes at every point in empty space. Of course, the field is singular (infinite) at any point containing a finite amount of mass, so we can't express the field due to a mass point precisely at the point. However, if we postulate a continuous distribution of gravitational charge (i.e., mass), with a density r_g specified at every point in a region, then it can be shown that the gravitational acceleration field at every point satisfies the equation

$$\nabla . g = -\rho_g \qquad ...(4)$$

Incidentally, if we define the gravitational potential (a scalar field) due to any particle of mass as φ = -m / r where r is the distance from the source particle (and noting that the potential due to multiple particles is simply additive), it's easy to show that

$$\nabla^2 \varphi = -\nabla . g$$

so equations (3) and (4) can be expressed equivalently in terms of the potential, in which case they are called Laplace's equation and Poisson's equation, respectively. The equation of motion for a test particle in the absence of any electromagnetic effects is simply a = g, so equation (2) gives the three components

$$\frac{d^2x}{dt^2} + \frac{m}{r^3}x = 0 \qquad \frac{d^2y}{dt^2} + \frac{m}{r^3}y = 0 \qquad \frac{d^2z}{dt^2} + \frac{m}{r^3}z = 0$$

To illustrate the use of these equations of motion, consider a circular path for our test particle, given by

x(t) = r sin(ωt) y(t) = r cos (ωt) z(t) = 0

In this case we see that r is constant and the second derivatives of x and y are $-r\omega^2\sin(\omega t)$ and $-r\omega^2\cos(\omega t)$ respectively. The equation of motion for z is identically satisfied and the equations for x and y both reduce to $r^3\omega^2 = m$, which is Kepler's third law for circular orbits.

Newton's analysis of gravity into a vectorial force field and a response was spectacularly successful in quantifying the effects of gravity, and by the beginning of the 20th century this approach was able to account for nearly all astronomical phenomena in the solar system within the limits of observational accuracy (the only notable exception being a slightly anomalous precession in the orbit of the planet Mercury, as discussed in previous sections). Based on this success, it was natural that the other forces of nature would be formalized in a similar way.

The next two most obvious forces that apply to material bodies are the electric and magnetic forces, represented by the last two terms in equation (1a). If we imagine that all of space is filled with a mist of tiny electrical charges q_i with velocities v_i, then we can define the classical charge density ρ_e and current density j as follows

$$\rho_e = \lim_{\Delta V \to 0} \frac{1}{\Delta V} \sum_{i \in \Delta V} q_i \qquad j = \lim_{\Delta V \to 0} \frac{1}{\Delta V} \sum_{i \in \Delta V} q_i V_i$$

where ΔV is an incremental volume of space. For the remainder of this section we will omit the subscript "e" with the understanding the r signifies the electric charge density. If we let x,y,z denote the position of the incremental quantity of charge, we can write out the individual components of the current density as

$$J_x = \rho \frac{dx}{dt} \qquad J_y = \rho \frac{dy}{dt} \qquad J_z = \rho \frac{dz}{dt}$$

Maxwell's equations for the electro-magnetic fields are

$$\nabla . E = \rho \qquad \text{...(5a)}$$

$$\nabla . B = 0 \qquad \text{...(5b)}$$

$$\nabla \times E = -\frac{\partial B}{\partial t} \qquad \text{...(5c)}$$

$$\nabla \times B = -\frac{\partial E}{\partial t} + j \qquad \text{...(5d)}$$

where E is the electric field, B is the magnetic field. Equations (5a) and (5b) suggest that the electric and magnetic fields are similar to the gravitational field g, since the divergences at each point equal the respective charge densities, with the difference being that the electric charge density may be positive or negative, and there does not exist (as far as we know) an isolated magnetic charge, i.e., no magnetic monopoles. Equations (5a) and (5b) are both static equations, in the sense that they do not involve the time parameter. By themselves they could be taken to indicate that the electric and magnetic fields are each individually similar to Newton's conception of the gravitational field, i.e., instantaneous "force-at-a-distance". (On this static basis we would presumably never have identified the magnetic field at all, assuming magnetic monopoles don't exist, and that the universe is not subject to any boundary conditions that caused B to be non-zero.)

However, equations (5c) and (5d) reveal a completely different aspect of the E and B fields, namely, that they are dynamically linked together, so the fields are not only functions of each other, but their definitions explicitly involve changes in time. Recall that the Newtonian gravitational field g was defined totally by the instantaneous spatial condition expressed by $\nabla \times g = -\rho_g$, so at any given instant the Newtonian gravitational field is totally determined by the spatial distribution of mass in that instant, consistent with the notion that simultaneity is absolute. In contrast, Maxwell's equations indicate that the fields E and B depend not only on the distribution of charge at a given putative "instant", but also on the movement of charge (i.e., the current density) and on the rates of change of the fields themselves at that "instant".

Since these equations contain a mixture of partial derivatives of the fields E and B with respect to the temporal as well as the spatial coordinates, dimensional consistency requires that the effective units of space and time must have a fixed relation to each other, assuming the units of E and B have a fixed relation. Specifically, the ratio of space units to time units must equal the ratio of electrostatic and electromagnetic units (all with respect to any frame of reference in which the above equations are applicable). This is the reason we were able to write the above equations without constant coefficients, because the fixed absolute ratio between the effective units of measure of time and space enables us to specify all the variables x,y,z,t in the same units.

Furthermore, this fixed ratio of space to time units has an extremely important physical significance for electromagnetic fields in empty space, where r and j are both zero. To see this, take the curl of both sides of (5c), which gives

$$\nabla \times \nabla \times E = -\nabla \times \frac{\partial B}{\partial t}$$

Now, for any arbitrary vector S it's easy to verify the identity

$$\nabla \times (\nabla \times S) = \nabla(\nabla . S) - \nabla^2 S$$

Therefore, we can apply this to the left hand side of the preceding equation, and noting that Ñ × E = 0 in empty space, we are left with

$$\nabla^2 E = \nabla \times \frac{\partial B}{\partial t}$$

Also, recall that the order of partial differentiation with respect to two parameters doesn't matter, so we can re-write the right-hand side of the above expression as

$$\nabla^2 E = \frac{\partial}{\partial t}(\nabla \times B)$$

Finally, since (5d) gives $\nabla \times B = \partial E/\partial t$ in empty space, the above equation becomes

$$\nabla^2 E = \frac{\partial^2 E}{\partial t^2} \qquad \text{...(6a)}$$

Similarly we can show that

$$\nabla^2 B = \frac{\partial^2 B}{\partial t^2} \qquad ...(6b)$$

Equations (6a) and (6b) are just the classical wave equation, which implies that electromagnetic changes propagate through empty space at a speed of 1 when using consistent units of space and time. In terms of conventional units this must equal the ratio of the electrostatic and electromagnetic units, which gives the speed

$$C = \frac{1}{\sqrt{\mu_0 \varepsilon_0}} \qquad ...(7)$$

where m_0 and e_0 are the permeability and permittivity of the vacuum. To some extent our choice of units is arbitrary, and in fact we conventionally define our units so that the permeability constant has the value

$\mu_0 = 4\pi \times 10^{-7}$ (kilogram× metre) / (ampere^2× second^2)

Since force has units of kg × m/sec^2 and charge has units of amp× sec, these conventions determine our units of force and charge, as well as distance, so we can then (theoretically) use Coulomb's law $F = q_1q_2/(4\pi\ \varepsilon_0\ r^2)$ to determine the permittivity constant by measuring the static force that exists between known electric charges at a certain distance. The best experimental value is

$\varepsilon_0 = 8.854187818 \times 10^{-12}$ (ampere^2 × second^4) / (kilogram× meter^3)

Substituting these values into equation (7) gives

$c = 2.997924579935 \times 10^8$ metre/second

This constant of proportionality between the units of space and time is based entirely on electrostatic and electromagnetic measurements, and it follows from Maxwell's equations that electromagnetic waves propagate at the speed c in a vacuum. (it was discussed in the history of attempts to measure the speed of light (which of course for most of human history was not known to be an electromagnetic phenomenon), but suffice it to say here that the best measured value for the speed of light is 299792457.4 m/sec, which agrees with Maxwell's predicted propagation speed for electromagnetic waves to nine significant digits.

This was Maxwell's greatest triumph, showing that electromagnetic waves propagate at the speed of light, from which we infer that light itself consists of electromagnetic waves, thereby unifying optics and electromagnetism. However, this magnificent result also presented Maxwell, and other physicists of the late 19th century, with a puzzle that would baffle them for decades. Equation (7) implies that, assuming the permittivity and permeability of the vacuum are the same when evaluated at rest with respect to any inertial frame of reference, in accord with the classical principle of relativity, and assuming Maxwell's equations are strictly valid in all inertial frames of reference, then it follows that the speed of light must be independent of the frame of reference. This agrees with the Galilean principle of relativity, but flatly violates the Galilean transformation rules, because it does not yield simply additive composition of speeds.

This was the conflict that vexed the young Einstein (age 16) when he was attending "prep school" in Aarau, Switzerland in 1895, preparing to re-take the entrance examination at the Zurich Polytechnic. Although he was deficient in the cultural subjects, he already knew enough mathematics and physics to realize that Maxwell's equations don't support the existence of a free wave at any speed other than c, which should be a fixed constant of nature according to the classical principle of relativity. But to admit an invariant speed seemed impossible to reconcile with the classical transformation rules.

Writing out equations (5d) and (5a) explicitly, we have four partial differential equations

$$\frac{\partial B_z}{\partial y} - \frac{\partial B_y}{\partial z} - \frac{\partial E_x}{\partial t} = \rho \frac{dx}{dt}$$

$$\frac{\partial B_x}{\partial z} - \frac{\partial B_z}{\partial x} - \frac{\partial E_y}{\partial t} = \rho \frac{dy}{dt}$$

$$\frac{\partial B_y}{\partial x} - \frac{\partial B_x}{\partial y} - \frac{\partial E_z}{\partial t} = \rho \frac{dz}{dt}$$

$$\frac{\partial E_x}{\partial x} - \frac{\partial E_y}{\partial y} - \frac{\partial E_z}{\partial z} = \rho \frac{dt}{dt}$$

The above equations strongly suggest that the three components of the current density j and the charge density ρ ought to be combined into a single four-vector, such that each component is the incremental charge per volume multiplied by the respective component of the four-velocity of the charge, as shown below

$$J_x = \rho \frac{dx}{d\tau} \qquad J_y = \rho \frac{dy}{d\tau} \qquad J_z = \rho \frac{dz}{d\tau} \qquad J_t = \rho \frac{dt}{d\tau}$$

where the parameter t is the proper time of the charge's rest frame. If the charge is stationary with respect to these x,y,z,t coordinates, then obviously the current density components vanish, and j_t is simply our original charge density ρ. On the other hand, if the charge is moving with respect to the x,y,z,t coordinates, we acquire a non-vanishing current density, and we find that the charge density is modified by the ratio dt/dt. However, it's worth noting that the incremental volume elements with respect to a moving frame of reference are also modified by the same Lorentz transformation, which ensures that the electrical charge on a physical object is invariant for all frames of reference.

We can also see from the four differential equations above that if the arguments of the partial derivatives on the left-hand side are arranged according to their denominators, they constitute a perfect anti-symmetric matrix

$$P = \begin{bmatrix} 0 & B_z & -B_y & -E_x \\ -B_z & 0 & B_x & -E_y \\ B_y & -B_x & 0 & -E_z \\ E_x & E_y & E_z & 0 \end{bmatrix}$$

If we let x^1, x^2, x^3, x^4 denote the coordinates x,y,z,t respectively, then equations (5a) and (5d) can be combined and expressed in the form

$$\sum_{\mu=1}^{4} \frac{\partial P_{\mu\nu}}{\partial X^{\mu}} = \rho \frac{dX^{\nu}}{d\tau} \qquad \text{...(8a)}$$

In exactly the same way we can combine equations (5b) and (5c) and express them in the form

$$\sum_{\mu=1}^{4} \frac{\partial Q_{\mu\nu}}{\partial X^{\mu}} = 0 \qquad \text{...(8b)}$$

where the matrix Q is an anti-symmetric matrix defined by

$$Q = \begin{bmatrix} 0 & E_z & -E_y & B_x \\ -E_z & 0 & E_x & B_y \\ E_y & -E_x & 0 & B_z \\ -B_x & -B_y & -B_z & 0 \end{bmatrix}$$

Returning again to equation (1a), we see that in the absence of a gravitational field the force on a particle with q = m = 1 and velocity v at a point in space where the electric and magnetic field vectors are E and B is given by

$$f = E + v \times B$$

In component form this can be written as

$$f_x i + f_y j + f_z k = E_x i + E_y j + E_z k$$

$$+ (v_y B_z - v_z B_y) + (v_z B_x - v_x B_z) + (v_x B_y - v_y B_x) K$$

Consequently the components of the acceleration are

$$f_x = E_x + (v_y B_z - v_z B_y)$$

$$f_y = E_y + (v_z B_x - v_x B_z)$$

$$f_z = E_z + (v_x B_y - v_y B_x)$$

To simplify the expressions, suppose the velocity of the particle with respect to the original x,y,z,t coordinates is purely in the positive x direction, i.e., we have $v_y = v_z = 0$ and $v_x = v$. Then the force on the particle has the components

$$f_x = E_x \qquad f_y = E_y - vB_z \qquad f_z = E_z + vB_y$$

Now consider the same physical situation, but with respect to a system of inertial coordinates x′,y′,z′,t′ in terms of which the particle's velocity is zero. To the first approximation we expect that the components of force are the same when evaluated with respect to the primed coordinate system, and in fact by symmetry it's clear that $f_{x'} = f_x$. However, for the components perpendicular to the velocity, the symmetry of the situation allows to say only that (for any fixed speed v) $f_{y'} = kf_y$ and $f_{z'} = kf_{z'}$, where A is a constant that approaches 1 for small v. Hence the components of the electric field with respect to the primed and unprimed coordinate systems are related according to

$$E_{x'} = E_x \qquad E_{y'} = k(E_y - vB_z) \qquad E_{z'} = k(E_z + vB_y)$$

By symmetry we can also write down the reciprocal transformation, replacing v with -v, which gives

$$E_x = E_{x'} \qquad E_y = k(E_{y'} - vB_{z'}) \qquad E_z = k(E_{z'} + vB_{y'})$$

Notice that we've used the same factor k for both transformations, because to the first order we know k(v) is simply 1, suggesting that the dependence of k on v is of the second order, which makes it likely that k(v) is an even function, i.e., we assume k(v) = k(-v). Substituting the expression for $E_{y'}$ into the expression for E_y and solving the resulting equation for $B_{z'}$ gives

$$B_{z'} = k\left(B_z - \left[\frac{k^2 - 1}{k^2 v^2}\right] vE_y\right)$$

By the same token, substituting the expression for $E_{z'}$ into the expression for E_z and solving for $B_{y'}$ gives

$$B_{z'} = k\left(B_y + \left[\frac{k^2 - 1}{k^2 v^2}\right] vE_z\right)$$

These last two expressions should look familiar, because they are formally identical to the expression for the transformed time coordinate developed in previous section. Letting $\phi(v)$ denote the quantity in square brackets for any given v, the general transformation equations for the electric and magnetic field components perpendicular to the velocity are

$$E_{y'} = \frac{E_y - vB_z}{\sqrt{1-\phi(v)v^2}} \qquad E_{z'} = \frac{E_z + vB_y}{\sqrt{1-\phi(v)v^2}}$$

$$B_{y'} = \frac{B_y + \phi(v)vE_z}{\sqrt{1-\phi(v)v^2}} \qquad B_{z'} = \frac{B_z + \phi(v)vE_y}{\sqrt{1-\phi(v)v^2}}$$

Comparing these equations with equation with previous equation it should come as no surprise that the actual transformations for the components of the electric and magnetic field are given by setting $\phi(v) = 1$. Consequently we have the invariants

$$E_{y'}{}^2 - B_{z'}{}^2 = E_y{}^2 - B_z{}^2 \qquad E_{z'}{}^2 - B_{y'}{}^2 = E_z{}^2 - B_y{}^2$$

Naturally we expect the field components parallel to the velocity to exhibit the corresponding invariance, i.e., we expect that

$$E_{x'}{}^2 - B_{x'}{}^2 = E_x{}^2 - B_x{}^2$$

from which we infer the final transformation equation $B_{x'} = B_x$. So, the complete set of transformation equations for the electric and magnetic field components from one system of inertial coordinates to another (with a relative velocity v in the positive x direction) is

$$E_{x'} = E_x \qquad E_{y'} = \frac{E_y - vB_z}{\sqrt{1-v^2}} \qquad E_{z'} = \frac{E_z + vB_y}{\sqrt{1-v^2}}$$

$$B_{x'} = B_x \qquad B_{y'} = \frac{B_y + vE_z}{\sqrt{1-v^2}} \qquad B_{z'} = \frac{B_z - vE_y}{\sqrt{1-v^2}}$$

Just as the Lorentz transformation for space and time intervals shows that those intervals are the components of a unified space-time interval, these transformation equations show that the electric and magnetic fields are components of a unified electro-magnetic field. The decomposition of the electromagnetic field into electric and magnetic components depends on the frame of reference. From the invariants noted above we see that, letting E^2 and B^2 denote squared magnitudes of the electric and magnetic field vectors at a given point, the quantity $E^2 - B^2$ is invariant (as is the dot product $E \times B$), analogous to the invariant $X^2 - T^2$ for spacetime intervals. The combined electromagnetic field can be represented by the matrix P defined previously, which transforms as a tensor of rank 2 under Lorentz transformations. So too does the matrix Q, and since Maxwell's equations can be expressed in terms of P and Q (as shown by equations (8a) and (8b)), we see that Maxwell's equations are invariant under Lorentz transformations.

THE INERTIA OF ENERGY

One of the first and most famous examples of the remarkable heuristic power of Einstein's relativistic interpretation was the suggestion that energy and inertial mass are, in a fundamental sense, equivalent. The word "suggestion" is used advisedly, because mass-energy equivalence is not a logically necessary consequence of special relativity. In fact, when combined with the gravitational equivalence principle, it turns out that mass-energy equivalence is technically incompatible with special relativity, and this was one of Einstein's main motivations for developing the general theory. Thus, although Einstein was originally led to the idea of mass-energy equivalence by the special theory, this equivalence ultimately made it necessary to generalize the theory in order to accommodate gravitation consistently.

It should also be mentioned that some kind of equivalence between mass and energy had long been suspected by physicists, even prior to 1905. For example, Poincare had noted that if Galilean relativity was applied to electrodynamics, the equivalence of mass and energy follows. Also, Lorentz had attempted to describe the mass of an electron as a manifestation of electromagnetic energy. (It's interesting that while some people were trying to "explain" electromagnetism as a disturbance in a material medium, others were trying to explain material substances as manifestations of electromagnetism!) However, the fact that mass-energy equivalence emerges so naturally from Einstein's kinematics, applicable to all kinds of mass and energy (not just electrons and electro-magnetism), was mainly responsible for the recognition of this equivalence as a general and fundamental aspect of nature. We'll first give a brief verbal explanation of how this equivalence emerges from Einstein's kinematics, and then follow with a quantitative description.

The basic principle of Einstein's special relativity is that inertial measures of spatial and temporal intervals are such that the velocity of light with respect to those measures is invariant. It follows that relative velocities are not transitively additive from one reference frame to another, and, as a result, the acceleration of an object with respect to one inertial frame must differ from its acceleration with respect to another inertial frame. However, by symmetry, an impact force exerted by two objects (in one spatial dimension) upon each another is equal and opposite, regardless of their relative velocity. These simple considerations lead directly to the idea that inertia (as quantified by mass) is an attribute of energy.

Given an object O of mass m, initially at rest, we apply a force F to the object, giving it an acceleration of F/m. After a while the object has achieved some velocity v, and we continue to apply the constant force F. But now imagine another inertial observer, this one momentarily co-moving with the object at this instant with a velocity v. This other observer sees a stationary object O of mass m subject to a force F, so, on the assumption that the laws of physics are the same in all inertial frames, we know that he will see the object respond with an acceleration of F/m (just as we did). However, due to non-additivity of velocities, the acceleration with respect to our measures of time and space must now be different. Thus, even though we're still applying a force F to the object, its acceleration (relative to our frame) is no longer equal to F/m. In fact, it must be less, and this acceleration must go to zero as v approaches the speed of light. Thus the effective inertia of the object in the direction of its motion increases. During this experiment we can also integrate the force we exerted over the distance traveled by the object, and determine the amount of work (energy) that we imparted to the object in bringing it to the velocity v. With a little algebra we can show that the ratio of the amount of energy we put into the object to the amount by which the object's inertia (units of mass) increased is exactly c^2.

To show this quantitatively, suppose the origin of a system of inertial coordinates K_0 is moving with speed u_0 relative to another system of inertial coordinates K, and if a particle P is moving with speed u (in the same direction as u_0) with respect to the K_0 coordinates, then the speed of the particle relative to the K coordinates is given by the velocity composition law

$$v = \frac{u_0 + u}{1 + \frac{u_0 u}{c^2}}$$

Differentiating with respect to u gives

$$\frac{dv}{du} = \frac{1 - \frac{u_0^2}{c^2}}{\left(1 + \frac{u_0 u}{c^2}\right)^2}$$

Hence, at the instant when P is momentarily co-moving with the K_0 coordinates, we have

$$dv = \left(1 - \frac{v^2}{c^2}\right) du$$

If we let τ and t denote the time coordinates of K_0 and K respectively, then from the metric $(d\tau)^2 = c^2(dt)^2 - (dx)^2$ and the fact that $v^2 = (dx/dt)^2$ it follows that the incremental lapse of proper time dt along the worldline of P as it advances from t to t + dt is $d\tau = dt\sqrt{1-(v/c)^2}$, so we can divide the above expression by this quantity to give

$$\frac{dv}{dt} = \left(1 - \frac{v^2}{c^2}\right)^{3/2} \frac{du}{d\tau}$$

The quantity $a = dv/d\tau$ is the acceleration of P with respect to the K coordinates, whereas $a_0 = du / dt$ is the "rest acceleration" of P with respect to the K_0 coordinates (relative to which it is momentarily at rest). Now, by symmetry, a force F exerted between a particle at rest in K on the particle P at rest in K_0 must be of equal and opposite magnitude with respect to both frames of reference. Also, by definition, a force of magnitude F applied to a particle of "rest mass" m_0 will result in an acceleration of $a_0 = F/m_0$ with respect to the reference frame in which the particle is momentarily at rest. Therefore, using the preceding relation between the accelerations with respect to the K_0 and K coordinates, we

$$F = \frac{m_0 a}{\left(1 - \frac{v^2}{c^2}\right)^{3/2}} \qquad \text{...(1)}$$

The coefficient of "a" in this expression has sometimes been called the "longitudinal mass", because it represents the effective proportionality between force and acceleration along the direction of action. Now let us define two quantities, p(v) and e(v), which we will call the momentum and kinetic energy of a particle of mass m_0 at any relative speed v. These quantities are defined respectively by the integrals of Fdt and Fds over an interval in which the particle is accelerated by a force F from rest to velocity v. The results of these integrations are independent of the pattern of acceleration, so we can assume constant acceleration "a" throughout the interval. Hence the integral of Fdt is evaluated from t = 0 to t = v/a, and since $s = (1/2)at^2$, the integral of Fds is evaluated from s = 0 to $s = v^2/(2a)$. In addition, we will define the inertial mass m of the particle as the ratio p/v. Therefore, the inertial mass and the kinetic energy of the particle at any speed v are given by

$$m = \frac{1}{v}\int_0^{v/a} F dt \qquad E = \int_C^{v^2/2a} F ds$$

If the force F were equal to $m_0 a$ (as in Newtonian mechanics) these two quantities would equal m_0 and $(1/2)m_0 v^2$ respectively. However, we've seen that consistency with relativistic kinematics requires the force to be given by equation (1). As a result, the inertial mass is given by $m = m_0/\sqrt{1-(v/c)^2}$, so it exceeds the rest mass whenever the particle has non-zero velocity. This increase in inertial mass is exactly proportional to the kinetic energy of the particle, as shown by

$$\frac{E}{m - m_0} = \frac{\int_0^{v^2/2a} F ds}{\frac{1}{v}\int_0^{v/a} F dt - m_0} = c^2$$

The exact proportionality between the extra inertia and the extra energy of a moving particle naturally suggests that it is the energy itself which has contributed the inertia, and this in turn suggests that all of the particle's inertia (including its rest inertia m_0) corresponds to some form of energy. This leads us to hypothesize a very general and important relation, $E = mc^2$, which signifies a fundamental equivalence between energy and inertial mass. From this we might imagine that all inertia is potentially convertible to energy, although it's worth noting that this does not follow rigorously from the principles of special relativity. It is just a hypothesis suggested by special relativity (as it is also suggested by Maxwell's equations). In 1905 the only experimental test that Einstein could imagine was to see if a lump of "radium salt" loses weight as it gives off radiation, but of course that would never be a complete test, because the radium doesn't decay down to nothing. The same is true with an nuclear bomb, i.e., it's really only the binding energy of the nucleus that is being converted, so it doesn't demonstrate an entire proton (for example) being converted into energy. However, today we can observe electrons and positrons annihilating each other completely, and yielding amounts of energy precisely in accord with the predictions of special relativity.

Incidentally, the above derivation followed Newton in adopting the Third Law (at least for impulse interactions along the line of motion) as a fundamental postulate, on the basis of symmetry. From this the conservation of momentum can be deduced. However, most modern treatments of relativity proceed in the opposite direction, postulating the conservation of momentum and then deducing something like the Third Law. (There are complications when applying the Third Law to extended interactions, and to interactions in which the forces are not parallel to the direction of motion, due to aberration effects and the ambiguity of simultaneity relations, but the preceding derivation was based solely on interactions that can be modeled as mutual contact events at single points, with the forces parallel to the direction of motion, in which case the Third Law is unproblematic.)

The typical modern approach to relativistic mechanics is to begin by defining momentum as the product of rest mass and velocity. One formal motivation for this definition is that the resulting 3-vector is well-behaved under Lorentz transformations, in the sense that if this quantity is conserved with respect to one inertial frame, it is automatically conserved with respect to all inertial frames (which would not be true if we defined momentum in terms of, say, longitudinal mass). On a more fundamental level, this definition is motivated by the fact that it agrees with non-relativistic momentum in the limit of low velocities. The heuristic technique of deducing the appropriate observable parameters of a theory from the requirement that they match classical observables in the classical limit was used extensively in early development of relativity, but apparently no one dignified the technique with a name until Bohr (characteristically) elevated it to the status of a "principle" in quantum mechanics, where it is known as the "Correspondence Principle".

Based on this definition, the modern approach then simply postulates that momentum is conserved. Then we define relativistic force as the rate of change of momentum. This is Newton's Second Law, and it's motivated largely by the fact that this "force", together with conservation of momentum, implies Newton's Third Law (at least in the case of contact forces).

However, from a purely relativistic standpoint, the definition of momentum as a 3-vector seems incomplete. It's three components are proportional to the derivatives of the three spatial coordinates x,y,z of the object with respect to the proper time τ of the object, but what about the coordinate time t? If we let x^j, j = 0, 1, 2, 3 denote the coordinates t,x,y,z, then it seems natural to consider the 4-vector

$$p^j = m\frac{dx^j}{d\tau} = m\frac{dt}{d\tau}\frac{dx^j}{dt} = \frac{m}{\sqrt{1-v^2}}\frac{dx^j}{dt}$$

where m is the rest mass. Then define the relativistic force 4-vector as the proper rate of change of momentum, i.e.,

$$f^j = \frac{dp^j}{d\tau} = m\frac{d^2x^j}{d\tau^2}$$

Our correspondence principle easily enables us to identify the three components p^1, p^2, p^3 as just our original momentum 3-vector, but now we have an additional component, p^0, equal to $m(dt/d\tau)$. Let's call this component the "energy" E of the object. In full four-dimensional spacetime coordinate time τ is related to the object's proper time τ according to

$$d\tau = \sqrt{(dt)^2 - (dx)^2 - (dy)^2 - (dz)^2}$$

$$= dt\sqrt{1 - \left[\left(\frac{dx}{dt}\right)^2 + \left(\frac{dy}{dt}\right)^2 + \left(\frac{dz}{dt}\right)^2\right]}$$

In geometric units (c = 1) the quantity in the square brackets is just v^2. Substituting back into our energy definition, we have

$$E = \frac{m}{\sqrt{1-v^2}} = m + \frac{1}{2}mv^2 + \frac{3}{8}mv^4 + \ldots \qquad \ldots(2)$$

The first term is simply m (or mc^2 in normal units), so we interpret this as the rest energy of the mass. This is sometimes presented as a derivation of mass-energy equivalence, but at best it's really just a suggestive heuristic device. The key step in this "derivation" was when we blithely decided to call p_0 the "energy" of the object. Strictly speaking, we violated our "correspondence principle" by making this definition, because by correspondence with the low-velocity limit, the energy E of a particle should be something like $(1/2)mv^2$, and clearly p_0 does not reduce to this in the low-speed limit. Nevertheless, we defined p_0 as the "energy" E, and since that component equals m when v = 0, we essentially just defined our result E = m (or $E = mc^2$ in ordinary units) for a mass at rest. From this reasoning it isn't clear that this is anything more than a bookkeeping convention, one that could just as well be applied in classical mechanics using some arbitrary squared velocity to convert from units of mass to units of energy. The assertion of physical equivalence between inertial mass and energy has significance only if it is actually possible for the entire mass of an object, including its rest mass, to manifestly exhibit the qualities of energy. Lacking this, the only equivalence between inertial mass and energy that special relativity strictly entails is the "extra" inertia that bodies exhibit when they acquire kinetic energy.

As mentioned above, even the fact that nuclear reactors give off huge amounts of energy does not really substantiate the complete equivalence of energy and inertial mass, because the energy given off in such reactions represents just the binding energy holding the nucleons (protons and neutrons) together. The binding energy is the amount of energy required to pull a nuclei apart. (The terminology is slightly inapt, because a configuration with high binding energy is actually a low energy configuration, and vice versa.) Of course, protons are all positively charged, so they repel each other by the Coulomb force, but at very small distances the strong nuclear force binds them together. Since each nucleon is attracted to every other nucleon, we might expect the total binding energy of a nucleus comprised of N nucleons to be proportional to N(N-1)/2, which would imply that the binding energy per nucleon would increase linearly with N. However, saturation effects cause the binding energy per nucleon to reach a maximum at for nuclei with $N \approx 60$ (e.g., iron), then to decrease slightly as N increases further. As a result, if an atom with (say) N = 230 is split into two atoms, each with N=115, the total binding energy per nucleon is increased, which means the resulting configuration is in a lower energy state than the original configuration.

In such circumstances, the two small atoms have slightly less total rest mass than the original large atom, but at the instant of the split the overall "mass-like" quality is conserved, because those two smaller atoms have enormous velocities, precisely such that the total relativistic mass is conserved. (This physical conservation is the main reason the old concept of relativistic mass has never been completely discarded.) If we then slow down those two smaller atoms by absorbing their energy, we end up with two atoms at rest, at which point a little bit of apparent rest mass has disappeared from the universe. On the other hand, it is also possible to fuse two light nuclei (e.g., N = 2) together to give a larger atom with more binding energy, in which case the rest mass of the resulting atom is less than the combined rest masses of the two original atoms. In either case (fission or fusion), a net reduction in rest mass occurs, accompanied by the appearance of an equivalent amount of kinetic energy and radiation. (The actual detailed mechanism by which binding energy, originally a "rest property" with isotropic inertia, becomes a kinetic property representing what we may call relativistic mass with anisotropic inertia, is not well understood.)

Another derivation of mass-energy equivalence is based on consideration of a bound "swarm" of particles, buzzing around with some average velocity. If the swarm is heated (i.e., energy E is added) the particles move faster and thereby gain both longitudinal and transverse mass, so the inertia of the individual particles is anisotropic, but since they are all buzzing around in random directions, the net effect on the stationary swarm (bound together by some unspecified means) is that its resistance to acceleration is isotropic, and its "rest mass" has effectively been increased by E/c^2. Of course, such a composite object still consists of elementary particles with some irreducible rest mass, so even this picture doesn't imply complete mass-energy equivalence.

To get complete equivalence we need to imagine something like photons bound together in a swarm. Now, it may appear that equation (2) fails to account for the energy of light, because it gives E proportional to the rest mass m, which is zero for a photon. However, the denominator of (2) is also zero for a photon (because v = 1), so the energy is indeterminate from that expression. However, we know from the study of electro-magnetic radiation that although a photon has no rest mass, it does (according to Maxwell's equations) have momentum, equal to $|p| = E$ (or E/c in conventional units). This suggests that we try to isolate the momentum component from the rest mass component of the energy. To do this, we square equation (2) and expand the simple geometric series as follows

$$E^2 = \frac{m^2}{1-v^2} = m^2 + m^2v^2 + m^2v^4 + m^2v^2 \ldots$$

Excluding the first term, which is purely rest mass, all the remaining terms are divisible by $(mv)^2$, so we can write this is

$$E^2 = m^2 + (mv)^2\left[1 + +v^2 + v^4 \ldots\right] = m^2 + \frac{(mv)^2}{1-v^2}$$

The right-most term is simply the squared magnitude of the momentum, so we have the apparently fundamental relation

$$E^2 = m^2 + |p|^2 \qquad ...(3)$$

Consistent with our premise that the E (or E/c in conventional units) equals the magnitude of the momentum |p| for a photon. Of course, electromagnetic waves are classically regarded as linear, meaning that photons don't ordinarily interfere with each other (directly). As Dirac said, "each photon interferes only with itself... interference between two different photons never occurs". However, the non-linear field equations of general relativity enable photons to interact gravitationally with each other. Wheeler coined the word "geon" to denote a swarm of massless particles bound together by the gravitational field associated with their energy, although he noted that such a configuration would be inherently unstable, viz., it would very rapidly either dissipate or shrink into complete gravitational collapse. Also, it's not clear that any physically realistic situation would lead to such a configuration in the first place, since it would require concentrating an amount of electromagnetic energy equivalent to the mass m within a radius of about $r = Gm/c^2$. For example, to make a geon from the energy equivalent of one electron, it would be necessary to concentrate that energy within a radius of about $(6.7)10^{-58}$ metres.

An interesting alternative approach to deducing (3) is based directly on the Minkowski metric

$$(d\tau)^2 = (dt)^2 - (dx)^2 - (dy)^2 - (dz)^2$$

This is applicable both to massive timelike particles and to light. In the case of light we know that the proper time dt and the rest mass m are both zero, but we may postulate that the ratio m/dt remains meaningful even when m and dt individually vanish. Multiplying both sides of the Minkowski line element by the square of this ratio gives immediately

$$m^2 = \left(\frac{dt}{d\tau}m\right)^2 - \left(\frac{dx}{d\tau}m\right)^2 - \left(\frac{dy}{d\tau}m\right)^2 - \left(\frac{dz}{d\tau}m\right)^2$$

The first term on the right side is E^2 and the remaining three terms are p_x^2, p_y^2, and p_z^2, so this equation can be written as

$$m^2 = E^2 - |p|^2$$

Hence this expression is nothing but the Minkowski spacetime metric multiplied through by $(m/dt)^2$, as illustrated in the figure below.

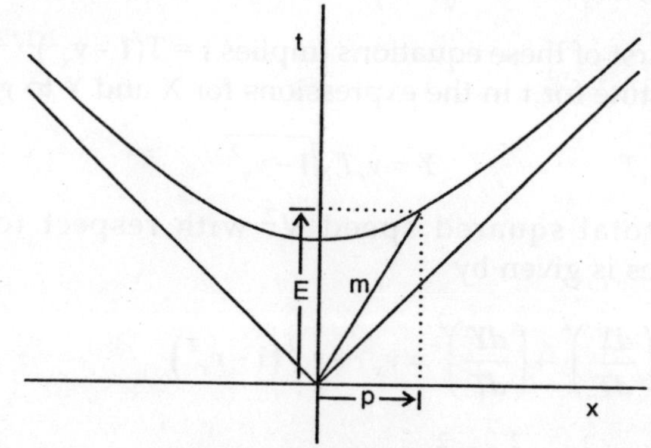

Fig. Inertia and Energy

The kinetic energy of the particle with rest mass m along the indicated worldline is represented in this figure by the portion of the total energy E in excess of the rest energy.

Returning to the question of how mass and energy can be regarded as different expressions of the same thing, recall that the energy of a particle with rest mass m_0 and speed V is $m_0/(1-V^2)^{1/2}$. We can also determine the energy of a particle whose motion is defined as the composition of two orthogonal speeds. Let t,x,y,z denote the inertial coordinates of system S, and let T,X,Y,Z denote the (aligned) inertial coordinates of system S′. In S the particle is moving with speed v_y in the positive y direction so its coordinates are

$$t = t \qquad x = 0 \qquad y = v_y t \qquad z = 0$$

The Lorentz transformation for a coordinate system S′ whose spatial origin is moving with the speed v in the positive x (and X) direction with respect to system S is

$$T = \frac{t - v_x X}{\sqrt{1 - v_x^{\,2}}} \qquad X = \frac{x - v_x t}{\sqrt{1 - v_x^{\,2}}} \qquad Y = y \qquad Z = z$$

so the coordinates of the particle with respect to the S′ system are

$$T = \frac{t}{\sqrt{1 - v_x^{\,2}}} \qquad X = \frac{vt}{\sqrt{1 - v_x^{\,2}}} \qquad Y = V_y t \qquad Z = 0$$

The first of these equations implies $t = T(1 - v_x^2)^{1/2}$, so we can substitute for t in the expressions for X and Y to give

$$X = v_x T \qquad Y = v_y T\sqrt{1 - v_x^{\,2}}$$

The total squared speed V^2 with respect to these coordinates is given by

$$V^2 = \left(\frac{dX}{dT}\right)^2 + \left(\frac{dY}{dT}\right)^2 = v_x^{\,2} + v_y^{\,2}\left(1 - v_x^{\,2}\right)$$

$$= v_x^{\,2} + v_y^{\,2} - v_x^{\,2} + v_y^{\,2}$$

Subtracting 1 from both sides and factoring the right hand side, this relativistic composition rule for orthogonal speeds v_x and v_y can be written in the form

$$1 - V^2 = \left(1 - v_x^{\,2}\right)\left(1 - v_y^{\,2}\right)$$

It follows that the total energy (neglecting stress and other forms of potential energy) of a ring of matter with a rest mass m_0 spinning with an intrinsic circumferential speed u and translating with a speed v in the axial direction is

$$E = \frac{m_0}{\sqrt{1 - V^2}} = \frac{m_0}{\sqrt{\left(1 - u^2\right)}\sqrt{\left(1 - v^2\right)}}$$

A similar argument applies to translatory motions of the ring in any direction, not just the axial direction. For example, consider motions in the plane of the ring, and focus on the contributions of two diametrically opposed particles (each of rest mass $m_0/2$) on the ring, as illustrated below.

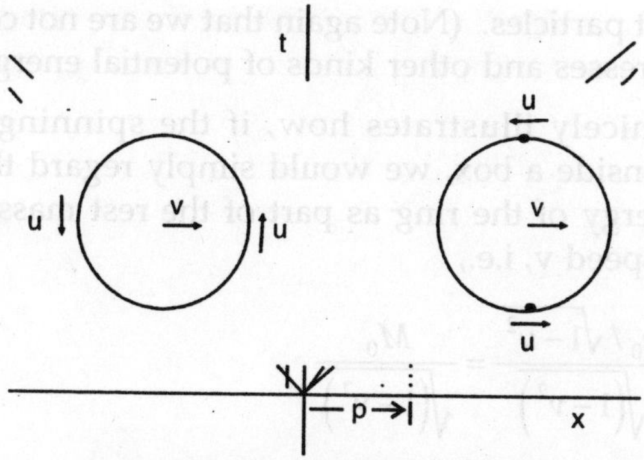

Fig. Inertia and Energy Equation

If the circumferential motion of the two particles happens to be perpendicular to the translatory motion of the ring, as shown in the left-hand figure, then the preceding formula for E is applicable, and represents the total energy of the two particles. If, on the other hand, the circumferential motion of the two particles is parallel to the motion of the ring's centre, as shown in the right-hand figure, then the two particles have the speeds (v+u)/(1+vu) and (v-u)/(1-vu) respectively, so the combined total energy (i.e., the relativistic mass) of the two particles is given by the sum

$$E = \frac{m_0/2}{\sqrt{1-\left(\frac{v+u}{1+vu}\right)^2}} + \frac{m_0/2}{\sqrt{1-\left(\frac{v-u}{1-vu}\right)^2}} = \frac{m_0}{\sqrt{1-u^2}\sqrt{1-v^2}}$$

Thus each pair of diametrically opposed particles with equal and opposite intrinsic motions parallel to the extrinsic translatory motion contribute the same total amount of energy as if their intrinsic motions were both perpendicular to the extrinsic motion. Every bound system of particles can be decomposed into pairs of particles with equal and opposite intrinsic motions, and these motions are either parallel or perpendicular or some combination relative to the extrinsic motion of the system, so the preceding analysis shows that the relativistic mass of the bound system of particles is isotropic, and the system behaves just like an object whose rest mass equals the sum of the intrinsic relativistic masses of the constituent particles. (Note again that we are not considering internal stresses and other kinds of potential energy.)

This nicely illustrates how, if the spinning ring was mounted inside a box, we would simply regard the angular kinetic energy of the ring as part of the rest mass M_0 of the box with speed v, i.e.,

$$E = \frac{m_0 / \sqrt{1-u^2}}{\sqrt{(1-v^2)}} = \frac{M_0}{\sqrt{(1-v^2)}}$$

where the "rest mass" of the box is now explicitly dependent on its energy content. This naturally leads to the idea that each original particle might also be regarded as a "box" whose contents are in an excited energy state via some kinetic mode (possibly rotational), and so the "rest mass" m_0 of the particle is actually just the relativistic mass of a lesser amount of "true" rest mass, leading to an infinite regress, and the idea that perhaps all matter is really some form of energy.

But does it really make sense to imagine that all the mass (i.e., inertial resistance) is really just energy, and that there is no irreducible rest mass at all? If there is no original kernel of irreducible matter, then what ultimately possesses the energy? Clearly Maxwell's equations can't provide an explanation for any bound state of pure electro-magnetic energy, because they are purely linear, whereas some non-linear interaction is required in order to yield a bound state. This is one of the main reasons that we know Maxwell's equations, by themselves, are not sufficient to account for the structure of the elementary particles of matter. Today we can observe the collision of an electron and a positron, two massive particles of matter, resulting in their mutual annihilation and the release of (apparently) pure energy in the form of photons, seeming to support the view that there is no irreducible "rest mass", and all matter is actually just a manifestation of energy.

To picture how an aggregate of massless energy can have non-zero rest mass, first consider two identical massive particles connected by a massless spring, as illustrated below.

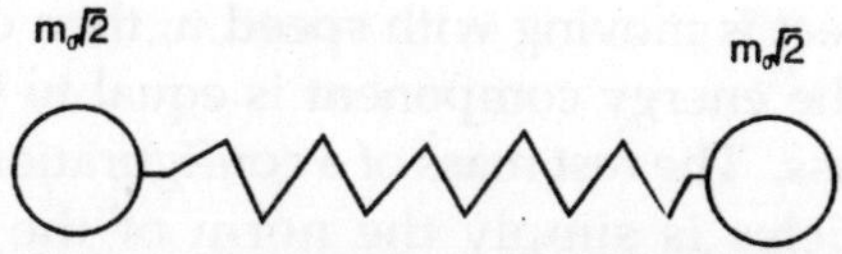

Suppose these particles are oscillating in a simple harmonic motion about their common centre of mass, alternately expanding and compressing the spring. The total energy of the system is conserved, but part of the energy oscillates between kinetic energy of the moving particles and potential (stress) energy of the spring. At the point in the cycle when the spring has no tension, the speed of the particles (relative to their common centre of mass) is a maximum. At this point the particles have equal and opposite speeds +u and -u, and we've seen that the combined rest mass of this configuration (corresponding to the amount of energy required to accelerate it to a given speed v) is $m_0/(1-u^2)^{1/2}$. At other points in the cycle, the particles are at rest with respect to their common centre of mass, but the total amount of energy in the system with respect to any given inertial frame is constant, so the effective rest mass of the configuration is constant over the entire cycle. Since the combined rest mass of the two particles themselves (at this point in the cycle) is just m_0, the additional rest mass to bring the total configuration up to $m_0/(1-u^2)^{1/2}$ must be contributed by the stress energy stored in the "massless" spring. This is one example of a massless entity acquiring rest mass by virtue of its stored energy.

Recall that the energy-momentum vector of a particle is defined as $[E, p_x, p_y, p_z]$ where E is the total energy and p_x, p_y, p_z are the components of the momentum, all with respect to some fixed system of inertial coordinates t,x,y,z. The rest mass m_0 of the particle is then defined as the Minkowskian "norm" of the energy-momentum vector, i.e.,

$$m_0^2 = E^2 - p_x^2 - p_y^2 - p_z^2$$

If the particle has rest mass m_0, then the components of its energy-momentum vector are

$$E = m_0 \frac{dt}{d\tau} \qquad p_x = m_0 \frac{dx}{d\tau} \qquad p_y = m_0 \frac{dy}{d\tau} \qquad p_z = m_0 \frac{dz}{d\tau}$$

If the object is moving with speed u, then $dt/d\tau = \gamma = 1/(1-u^2)^{1/2}$, so the energy component is equal to the transverse relativistic mass. The rest mass of a configuration of arbitrarily moving particles is simply the norm of the sum of their individual energy-momentum vectors. The energy-momentum vectors of two particles with individual rest masses m0 moving with speeds $dx/dt = u$ and $dx/dt = -u$ are $[\gamma m_0, \gamma m_0 u, 0, 0]$ and $[\gamma m_0, -\gamma m_0 u, 0, 0]$, so the sum is $[2gm_0, 0, 0, 0]$, which has the norm $2gm_0$. This is consistent with the previous result, i.e., the rest mass of two particles in equal and opposite motion about the centre of the configuration is simply the sum of their (transverse) relativistic masses, i.e., the sum of their energies.

A photon has no rest mass, which implies that the Minkowskian norm of its energy-momentum vector is zero. However, it does not follow that the components of its energy-momentum vector are all zero, because the Minkowskian norm is not positive-definite. For a photon we have $E^2 - p_x{}^2 - p_y{}^2 - p_z{}^2 = 0$ (where E = hn), so the energy-momentum vectors of two photons, one moving in the positive x direction and the other moving in the negative x direction, are of the form [E, E, 0, 0] and [E, -E, 0, 0] respectively. The Minkowski norms of each of these vectors individually are zero, but the sum of these two vectors is [2E, 0, 0, 0], which has a Minkowski norm of 2E. This shows that the rest mass of two identical photons moving in opposite directions is $m_0 = 2E = 2hn$, even though the individual photons have no rest mass.

If we could imagine a means of binding the two photons together, like the two particles attached to the massless spring, then we could conceive of a bound system with positive rest mass whose constituents have no rest mass. As mentioned previously, in normal circumstances photons do not interact with each other (i.e., they can be superimposed without affecting each other), but we can, in principle, imagine photons bound together by the gravitational field of their energy (geons). The ability of electrons and anti-electrons (positrons) to completely annihilate each other in a release of energy suggests that these actual massive particles are also, in some sense, bound states of pure energy, but the mechanisms or processes that hold an electron together, and that determine its characteristic mass, charge, etc., are not known.

It's worth noting that the definition of "rest mass" is somewhat context-dependent when applied to complex accelerating configurations of entities, because the momentum of such entities depends on the space and time scales on which they are evaluated. For example, we may ask whether the rest mass of a spinning disk should include the kinetic energy associated with its spin. For another example, if the Earth is considered over just a small portion of its orbit around the Sun, we can say that it has linear momentum (with respect to the Sun's inertial rest frame), so the energy of its circumferential motion is excluded from the definition of its rest mass. However, if the Earth is considered as a bound particle during many complete orbits around the Sun, it has no net momentum with respect to the Sun's frame, and in this context the Earth's orbital kinetic energy is included in its "rest mass".

Similarly a "stationary" chunk of lead not microscopically stationary, but in the aggregate, averaged over the characteristic time scale of the mean free oscillation time of a particle, it is stationary, and is treated as such. The temperature of the lead actually represents changes in the states of motion of the constituent particles, but over a suitable length of time the particles are still stationary. We can continue to smaller scales, down to sub-atomic particles comprising individual atoms, and we find that the position and momentum of a particle cannot even be precisely stipulated simultaneously. In each case we must choose a context in order to apply the definition of rest mass.

Physical entities possess multiple modes of excitation (kinetic energy), and some of these modes we may choose (or be forced) to absorb into the definition of the object's "rest mass", because they do not vanish with respect to any inertial reference frame, whereas other modes we may choose (and be able) to exclude from the "rest mass". In order to assess the momentum of complex physical entities in various states of excitation, we must first decide how finely to decompose the entities, and the time intervals over which to make the assessment. The "rest mass" of an entity invariably includes some of what would be called energy or "relativistic mass" if we were working on a lower level of detail.

DOPPLER SHIFT FOR SOUND AND LIGHT

For historical reasons, some older text books present two different versions of the Doppler shift equations, one for acoustic phenomena based on traditional Newtonian kinematics, and another for optical and electromagnetic phenomena based on relativistic kinematics. This sometimes gives the impression that relativity requires us to apply a different set of kinematical rules to the propagation of sound than to the propagation of light, but of course that is not the case. The kinematics of relativity apply uniformly to the propagation of all kinds of signals, provided we give the exact formulae. The traditional acoustic formulas are inexact, tacitly based on Newtonian approximations, but when they are expressed exactly we find that they are perfectly consistent with the relativistic formulas.

Consider a frame of reference in which the medium of signal propagation is assumed to be at rest, and suppose an emitter and absorber are located on the x axis, with the emitter moving to the left at a speed of v_e and the absorber moving to the right, directly away from the emitter, at a speed of v_a. Let c_s denote the speed at which the signal propagates with respect to the medium. Then, according to the classical (non-relativistic) treatment, the Doppler frequency shift is

$$V_a = \left(\frac{1 - v_a / c_s}{1 + v_e / C_s} \right) V_e$$

(It's assumed here that u and v are less than c_s, because otherwise there may be shock waves and/or lack of communication between transmitter and receiver, in which case the Doppler effect does not apply.) The above formula is often quoted as the Doppler effect for sound, and then another formula is given for light, suggesting that relativity arbitrarily treats sound and light signals differently. In truth, relativity has just a single formula for the Doppler shift, which applies equally to both sound and light. This formula can basically be read directly off the spacetime diagram shown below

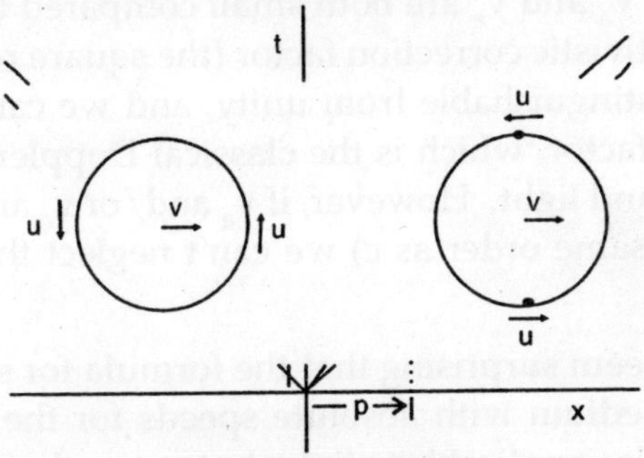

Fig. Doppeler Shift

If an emitter on worldline OA turns a signal ON at event O and OFF at event A, the proper duration of the signal is the magnitude of OA, and if the signal propagates with the speed of the worldline AB, then the proper duration of the pulse for a receiver on OB will equal the magnitude of OB. Thus we have

$$C_s = \frac{X_B - X_A}{t_B - t_A} \qquad V_e = -\frac{X_A}{t_A} \qquad V_a = \frac{X_B}{t_B}$$

and

$$|OA| = \sqrt{t_A^{\,2} - (X_A/C)^2} = t_A\sqrt{1-(X_e/C)^2}$$

$$|OB| = \sqrt{t_B^{\,2} - (X_B/C)^2} = t_B\sqrt{1-(X_a/C)^2}$$

Substituting $x_A = -v_e t_A$ and $x_B = v_a t_B$ into the equation for c_s and re-arranging terms gives

$$(t_B - t_A)C_s = V_a t_B + V_e t_A$$

from which we get $\dfrac{t_A}{t_B} = \dfrac{1 - v_a/C_s}{1 + v_e/C_s}$

Substituting this into the ratio of |OA| / |OB| gives the ratio of proper times for the signal, which is the inverse of the ratio of frequencies:

$$\frac{V_a}{V_e} = \frac{|OA|}{|OB|} = \frac{1 - v_a/C_s}{1 + v_a/C_s}\sqrt{\frac{1-(v_e/C)^2}{1-(v_a/C)^2}}$$

Now, if v_a and v_e are both small compared to c, it's clear that the relativistic correction factor (the square root quantity) will be indistinguishable from unity, and we can simply use the leading factor, which is the classical Doppler formula for both sound and light. However, if v_a and/or v_e are fairly large (i.e., on the same order as c) we can't neglect the relativistic correction.

It may seem surprising that the formula for sound waves in a fixed medium with absolute speeds for the emitter and absorber is also applicable to light, but notice that as the signal propagation speed c_s goes to c, the above Doppler formula smoothly evolves into

$$V_a = V_e\sqrt{\left(\frac{1 - v_a/c}{1 + v_a/c}\right)\left(\frac{1 - v_e/c}{1 + v_e/c}\right)}$$

which is very nice, because we immediately recognize the quantity inside the square root as the multiplicative form of the relativistic composition law for velocities. In other words, letting u denote the composition of the speeds v_a and v_e given by the formula

$$w = \frac{v_a + v_e}{1 + v_a v_e / c^2}$$

it follows that

$$\left(\frac{1 - u/c}{1 + u/c}\right) = \left(\frac{1 - v_a/c}{1 + v_a/c}\right)\left(\frac{1 - v_e/c}{1 + v_e/c}\right)$$

Consequently, as c_s increases to c, the absolute speeds v_e and v_a of the emitter and absorber relative to the fixed medium merge into a single relative speed u between the emitter and absorber, independent of any reference to a fixed medium, and we arrive at the relativistic Doppler formula for waves propagating at c for an emitter and absorber with a relative velocity of u:

$$V_a = V_e\sqrt{\frac{1-u/c}{1+u/c}}$$

To clarify the relation between the classical and relativistic Doppler shift equations, recall that for a classical treatment of a wave with characteristic speed c_s in a material medium the Doppler frequency shift depends on whether the emitter or the absorber is moving relative to the fixed medium. If the absorber is stationary and the emitter is receding at a speed of v (normalized so $c_s = 1$), then the frequency shift is given by

$$\frac{V_a}{V_e} = \frac{1}{1+v} = 1 - v + v^2 + \ldots$$

whereas if the emitter is stationary and the absorber is receding the frequency shift is

$$\frac{V_a}{V_e} = 1 - v$$

To the first order these are the same, but they obviously differ significantly if v is close to 1. In contrast, the relativistic Doppler shift for light, with $c_s = c$, does not distinguish between emitter and absorber motion, but simply predicts a frequency shift equal to the geometric mean of the two classical formulas, i.e.,

$$\frac{V_a}{V_e} = \sqrt{\frac{1-v}{1+v}} = 1 - v + \frac{1}{2}v^2 - \ldots$$

Naturally to first order this is the same as the classical Doppler formulas, but it differs from both of them in the second order, so we should be able to check for this difference, provided we can arrange for emitters and/or absorbers to be moving with significant speeds. The Doppler effect has in fact been tested at speeds high enough to distinguish between these two formulas. The possibility of such a test, based on observing the Doppler shift for "canal rays" emitted from high-speed ions, had been considered by Stark in 1906, and Einstein published a short paper in 1907 deriving the relativistic prediction for such an experiment. However, it wasn't until 1938 that the experiment was actually performed with enough precision to discern the second order effect. In that year, Ives and Stilwell shot hydrogen atoms down a tube, with velocities (relative to the lab) ranging from about 0.8 to 1.3 times 10^6 m/sec. As the hydrogen atoms were in flight they emitted light in all directions. Looking into the end of the tube (with the atoms coming toward them), Ives and Stilwell measured a prominent characteristic spectral line in the light coming forward from the hydrogen. This characteristic frequency n was Doppler shifted toward the blue by some amount $dn_{approach}$ because the source was approaching them. They also placed a mirror at the opposite end of the tube, behind the hydrogen atoms, so they could look at the same light from behind, i.e., as the source was effectively moving away from them, red-shifted by some amount dn_{receed}. The following is a table of results from the original 1938 experiment for four different velocities of the hydrogen atom:

$$\left(10^5\right)\left(dv_{approach} - dv_{receed}\right)/v$$

$v, 10^6\, m/s$	*Classical*	Re*lativistic*	*Experiment*
0.865	1.67	0.835	0.762
0.01	2.26	1.13	1.1
1.15	2.90	1.45	1.42
1.33	3.94	1.97	1.9

Ironically, although the results of their experiment brilliantly confirmed Einstein's prediction based on the special theory of relativity, Ives and Stillwell were not advocates of relativity, and in fact gave a completely different theoretical model to account for their experimental results and the deviation from the classical prediction. This illustrates the fact that the results of an experiment can never uniquely identify the explanation. They can only split the range of available models into two groups, those that are consistent with the results and those that aren't. In this case it's clear that any model yielding the classical prediction is ruled out, while the Lorentz/Einstein model is found to be consistent with the observed results.

All the above was based on the assumption that the emitter and absorber are moving relative to each other directly along their "line of sight". More generally, we can give the Doppler shift for the case when the (inertial) motions of the emitter and absorber are at any specified angles relative to the "line of sight". Without loss of generality we can assume the absorber is stationary at the origin of inertial coordinates and the emitter is moving at a speed v and at an angle f relative to the direct line of sight, as illustrated below.

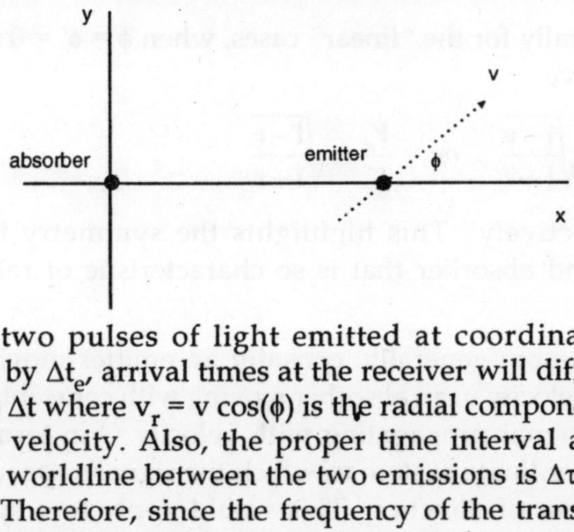

For two pulses of light emitted at coordinate times differing by Δt_e, arrival times at the receiver will differ by $\Delta t_a = (1 + v_r)\,\Delta t$ where $v_r = v\cos(\phi)$ is the radial component of the emitter's velocity. Also, the proper time interval along the emitter's worldline between the two emissions is $\Delta\tau_e = \Delta t_e\,(1 - v^2)^{1/2}$. Therefore, since the frequency of the transmissions with respect to the emitter's rest frame is proportional to 1/ Dt_e, and the frequency of receptions with respect to the absorber's rest frame is proportional to $1/\Delta\tau_a$, the full frequency shift is

$$V_a = \frac{\sqrt{1-v^2}}{1+v\cos(\phi)}V_e$$

This differs in appearance from the Doppler shift equation given in Einstein's 1905 paper, but only because, in Einstein's equation, the angle ϕ is evaluated with respect to the emitter's rest frame, whereas in our equation the angle is evaluated with respect to the absorber's rest frame. These two angles differ because of the effect of aberration. If we let ϕ' denote the angle with respect to the emitter's rest frame, then ϕ' is related to ϕ by the aberration equation

$$\cos(\phi) = \frac{\cos(\phi') - v}{1 - v\cos(\phi')}$$

Substituting for $\cos(\phi)$ into the previous equation gives Einstein's equation for the Doppler shift, i.e.,

$$V_a = \frac{1 - v\cos(\phi')}{\sqrt{1 - v^2}} V_e$$

Naturally for the "linear" cases, when $\phi = \phi' = 0$ or $\phi = \phi' = \pi$ we have

$$\frac{V_a}{V_e} = \sqrt{\frac{1-v}{1+v}} \quad or \quad \frac{V_a}{V_e} = \sqrt{\frac{1-v}{1-v}}$$

respectively. This highlights the symmetry between emitter and absorber that is so characteristic of relativistic physics.

Even more generally, consider an emitter moving with constant velocity u, an absorber moving with constant velocity v, and a signal propagating with velocity C in terms of an inertial coordinate system in which the signal's speed $|C|$ is independent of direction. This would apply to a system of coordinates at rest with respect to the medium of the signal, and it would apply to any inertial coordinate system if the signal is light in a vacuum. It would also apply to the case of a signal emitted at a fixed speed relative to the emitter, but only if we take u = 0, because in this case the speed of the signal is independent of direction only in terms of the rest frame of the emitter. We immediately have the relation

$$|r_a - r_e|^2 = |C|^2 (t_a - t_e)^2$$

Where r_e and r_a are the position vectors of the emission and absorption events at the times t_e and t_a respectively. Differentiating both sides with respect to t_a and dividing through by $2(t_a - t_e)$, and noting that $(r_a - r_e)/(t_a - t_e) = C$, we get

$$C.\left(v - u\frac{dt_e}{dt_a}\right) = |C|^2\left(1 - \frac{dt_e}{dt_a}\right)$$

where u and v are the velocity vectors of the emitter and absorber respectively. Solving for the ratio dt_e/dt_a, we arrive at the relation

$$\frac{dt_e}{dt_a} = \frac{|C|^2 - C.v}{|C|^2 - C.u} = \frac{C./(C-v)}{C.(C-u)} = \frac{1 - \dfrac{C.v}{|C|^2}}{1 - \dfrac{C.u}{|C|^2}}$$

Making use of the dot product identity $r.s = |r|\,|s|\cos(\theta_{r,s})$ where $\theta_{r,s}$ is the angle between the r and s vectors, these can be re-written as

$$\frac{dt_e}{dt_a} = \frac{1 - \dfrac{|v|}{|C|}\cos\left(\theta_{c,v}\right)}{1 - \dfrac{|u|}{|C|}\cos\left(\theta_{c,u}\right)} = \frac{|C-v|\cos\left(\theta_{C,C-v}\right)}{|C-u|\cos\left(\theta_{C,C-u}\right)}$$

The frequency of any process is inversely proportional to the duration of the period, so the frequency at the absorber relative to the emitter, projected by means of the signal, is given by $n_a/n_e = dt_e/dt_a$. Therefore, the above expressions represent the classical Doppler effect for arbitrarily moving emitter and receiver. However, the elapsed proper time along a worldline moving with speed v in terms of any given inertial coordinate system differs from the elapsed coordinate time by the factor

$$\frac{d\tau}{dt} = \sqrt{1 - (v/c)^2}$$

Where c is the speed of light in vacuum. Consequently, the actual ratio of proper times – and therefore proper frequencies – for the emitter and absorber is

$$\frac{V_a}{V_e}=\frac{d\tau_e}{d\tau_a}=\frac{1-\frac{|v|}{|C|}\cos\left(\theta_{c,v}\right)}{1-\frac{|u|}{|C|}\cos\left(\theta_{c,u}\right)}\sqrt{\frac{1-(u/c)^2}{1-(v/c)^2}}$$

The leading ratio is the classical Doppler effect, and the square root factor is the relativistic correction.

Chapter 4

Acceleration and Gravity

ACCELERATION AND GENERAL RELATIVITY

General relativity is Einstein's extension of special relativity to include gravity. An important aspect of general relativity is that spacetime is no longer necessarily flat, but in fact may be curved under the influence of mass. Understanding curved spacetime is an advanced topic which is not easily accessible at the level of this course. However, it turns out that some insight into general relativistic phenomena may be obtained by investigating the effects of acceleration in the flat (but non-Euclidean) space of special relativity.

When an object's velocity changes, it accelerates. Acceleration shows the change in velocity in a unit time. Velocity is measured in metres per second, m/s, so acceleration is measured in (m/s)/s, or m/s^2, which can be both positive and negative.

The central assumption of general relativity is the equivalence principle, which states that gravity is a force which arises from being in an accelerated reference frame. To understand this we must first investigate the concept of acceleration. We then see how this leads to phenomena such as the gravitational red shift, event horizons, and black holes. We also introduce in a preliminary way the notions of force and mass.

ACCELERATION

Imagine that you are in a powerful luxury car stopped at a stoplight. As you sit there, gravity pushes you into the comfortable leather seat. The light turns green and you "floor it". The car accelerates and and an additional force pushes you into the seat back. You round a curve, and yet another force pushes you toward the outside of the curve. (But the well designed seat and seat belt keep you from feeling discomfort!)

The average acceleration is the ratio between the change in velocity and the time interval.

$$a = \frac{\Delta v}{\Delta t} = \frac{Velocity\,Change}{Ellapsed\,Time}$$

For example, if a car moves from the rest to 5 m/s in 5 seconds, its average acceleration is

$$a = \frac{5m/s}{5s} = 1m/s^2$$

An instantaneous acceleration is the change in velocity at one moment. When the velocity of an object changes it is said to be accelerating or more formally acceleration is the rate of change of velocity with time.

In everyday English, the word acceleration is often used to describe a state of increasing speed. For many Americans, their only experience with acceleration comes from car ads. When a commercial shouts "zero to sixty in six point seven seconds" what they're saying here is that this particular car takes 6.7 s to reach a speed of 60 mph starting from a complete stop. This example illustrates acceleration as it is commonly understood, but acceleration in physics is much more than just increasing speed.

Any change in the velocity of an object results in an acceleration: increasing speed (what people usually mean when they say acceleration), decreasing speed (also called deceleration or retardation), or changing direction. Yes, that's right, a change in the direction of motion results in an acceleration even if the speed didn't change. That's because acceleration depends on a change in velocity and velocity is a vector quantity — one with both magnitude and direction. Thus, a falling apple accelerates, a car stopping at a traffic light accelerates, and an orbiting planet accelerates. Acceleration occurs anytime an object's speed increases, speed decreases, or direction of motion changes.

As was the case with velocity, there are two kinds of acceleration – average and instantaneous. Average acceleration is measured over a "long" (measurable) time interval while instantaneous acceleration is measured over a "very small" (infinitesimal) time interval. The formulas are also similar to those of velocity. Acceleration is defined by the following equations ...

$$\bar{a} = \frac{\Delta v}{\Delta t} \quad \text{average acceleration}$$

$$a = \lim_{\Delta t \to 0} \frac{\Delta v}{\Delta t} = \frac{dv}{dt} = \frac{d^2 r}{dt^2} \quad \text{instantaneous acceleration}$$

For those of you familiar with calculus, check out the second equation, which states that acceleration is the first derivative of velocity with respect to time and the second derivative of displacement with respect to time. Or if you prefer, acceleration is the rate of change in velocity and also (since velocity is a change in displacement) the rate of change of the rate of change of displacement.

Units

Calculating acceleration involves dividing velocity by time – or in terms of units, dividing metres per second [m/s] by second [s]. Dividing distance by time twice is the same as dividing distance by the square of time. The SI unit of acceleration is the metre per second squared [m/s^2].

Another frequently used unit is the acceleration due to gravity. Since we are all familiar with the effects of gravity on ourselves and the objects around us it makes for a convenient standard against which we can compare other accelerations. Everything feels normal at 1 g, twice as heavy at 2 g, and weightless at 0 g. This unit has a very precise definition, but for everyday use 9.8 m/s^2 is sufficient.

The unit called the acceleration due to gravity, with the symbol g (roman), is not the same as the natural phenomena (discussed in a later section) called the acceleration due to gravity, having the symbol g (italic). The former is a constant whereas the latter is a variable.

Although the term "g force" is often used, the g is a measure of acceleration and not force.

- The acceleration due to gravity ...
  - is a natural unit of acceleration,
  - is represented by the symbol g (roman),
  - is equal to 9.80665 m/s^2 by definition,
  - is often rounded to 9.8 m/s^2 for convenience, and
  - is sometimes called the "g force" even though it is not a measure of force.

Of particular concern to humans are the physiological effects of acceleration. To put things in perspective, all values are stated in g.

- In roller coaster design, speed is of the essence. Or, is it? If speed was all there was to designing a thrill ride, then the freeway would be pretty exciting. Most roller coaster rarely exceed 30 m/s (60 mph). Contrary to popular belief, it is the acceleration that makes the ride. A well designed roller coaster will subject the rider to maximum accelerations of 3 to 4 g for brief periods. This gives the ride a faster, more dangerous feel.
- Despite the immense power of its engines, the acceleration of the Space Shuttle is kept below 3 g. Anything greater would put unnecessary stress on the astronauts and the ship itself. Once in orbit, they enter an extended period of free fall, which provides the sensation of weightlessness. Such a "zero g" environment can also be simulated inside a specially piloted aircraft or a free fall drop tower.
- Fighter pilots can experience accelerations of up to 8 g for brief periods during tactical maneuvers. If sustained for more than a few seconds, 4 to 6 g is

sufficient to induce blackout. To prevent loss of consciousness, fighter pilots wear special pressure suits that squeeze the legs and abdomen, forcing blood to remain in the head.

- Pilots and astronauts often train in human centrifuges capable of up to 15 g. Exposure to such intense accelerations is kept very brief for safety reasons. Humans are rarely subjected to anything higher than 8 g for longer than a few seconds.
- Acceleration is related to injury. This is why the most common sensor in a crash test dummy is the accelerometer. Extreme acceleration can lead to death,and cause many deaths in road accidents evry year. The acceleration during the crash that killed Diana, Princess of Wales, in an accident was estimated to have been on the order of 70 to 100 g, which was intense enough to tear the pulmonary artery from her heart – an injury that is nearly impossible to survive. Had she been wearing a seat belt, the acceleration would have been something more like 30 or 35 g – enough to break a rib or two, but not nearly enough to kill most people.

Summary

- Acceleration is the rate of change of velocity with time.
  - As a vector it must be stated with both magnitude and direction.
- Acceleration occurs anytime an object's ...
  - speed increases,
  - speed decreases, or
  - direction of motion changes.
- Average acceleration ...
  - is measured over a non-zero time interval and
  - is represented by the symbol a_{ave} or a (overline)

- Instantaneous acceleration ...
  - is the limit of average acceleration as the time interval approaches zero,
  - is the first derivative of velocity with respect to time,
  - is the second derivative of displacement with respect to time, and
  - is represented by the symbol a (boldface)
- The various forms of acceleration are defined by the following equations ...

$$\bar{a} = \frac{\Delta v}{\Delta t} \quad \text{average acceleration}$$

$$a = \lim_{at \to 0} \frac{\Delta v}{\Delta t} = \frac{dv}{dt} = \frac{d^2 r}{dt^2} \quad \text{instantaneous acceleration}$$

- The SI unit of acceleration is the metre perv second squared [m/s^2].
- The acceleration due to gravity ...
  - is a natural unit of acceleration,
  - is represented by the symbol g (roman),
  - is equal to 9.80665 m/s^2 by definition,
  - is often rounded to 9.8 m/s^2 for convenience, and
  - is sometimes called the "g force" even though it is not a measure of force.

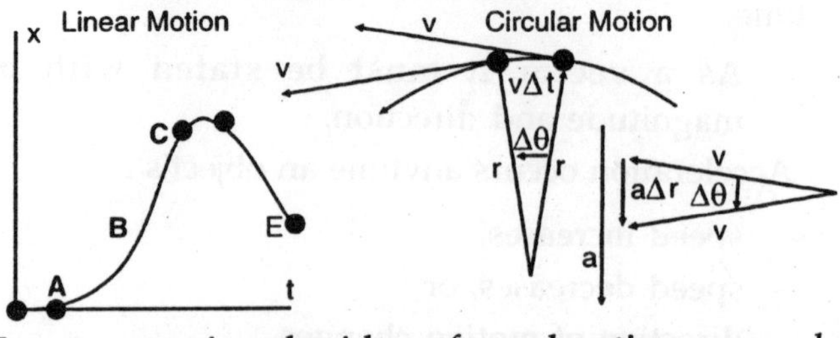

Let us examine the idea of acceleration more closely. Considering first acceleration in one dimension, the left panel of figure shows the position of an object as a function of time, $x(t)$. The velocity is simply the time rate of change of the position:

$$v(t) = \frac{dx(t)}{dt},$$

The acceleration is the time rate of change of velocity:

$$a(t) = \frac{dv(t)}{dt} = \frac{d^2x(t)}{dt^2},$$

In the left panel of figure above only the segment OA has zero velocity. Velocity is increasing in AB, and the acceleration is positive there. Velocity is constant in BC, which means that the acceleration is zero. Velocity is decreasing in CD, and the acceleration is negative. Finally, in DE, the velocity is negative and the acceleration is zero.

In two or three dimensions, position *r*, velocity *v*, and acceleration *a* are all vectors, so that the velocity is

$$v(t) = \frac{dr(t)}{dt},$$

while the acceleration is

$$a(t) = \frac{dv(t)}{dt},$$

Thus, over some short time interval Δt the changes in and can be written

$$\Delta r = v\Delta t \qquad \Delta v = a\Delta t;$$

These are vector equations, so the subtractions implied by the "delta" operations must be done vectorially. An example where the vector nature of these quantities is important is motion in a circle at constant speed.

CIRCULAR MOTION

Imagine an object being swung around in a circle on a string, as shown in figures. The right panel in above figure shows the position of the object at two times spaced by the time interval Δt. The position vector of the object relative to the centre of the circle rotates through an angle $\Delta\theta$ during this interval, so the angular rate of revolution of the object about the centre is ω = Δθ/Δ*t*. The magnitude of the velocity of the object is *v*, so the object moves a distance *v*Δ*t* during the time interval. To the extent that this distance is small compared to the radius *r* of the circle, the angle Δθ = Δ*t*/*r*. Solving for and using ω = Δθ/Δ*t*, we see that

v = ωr (curcular motion)

The direction of the velocity vector changes over this interval, even though the magnitude *v* stays the same. This motion shows that this change in direction implies an acceleration *a* which is directed toward the centre of the circle. The magnitude of the vectoral change in velocity in the time interval Δ*t* is *a*Δ*t*. Since the angle between the initial and final velocities is the same as the angle $\Delta\theta$ between the initial and final radius vectors, we see from the geometry of the triangle in the right sub-panel of circular figure shows. *a*Δ*t*/*v* = $\Delta\theta$. Solving for results in

a =ω*v* (circular motion)

Combining equations yields the equation for centripetal acceleration, i. e., the acceleration toward the centre of a circle:

$a=\omega^2 r = v^2/r$ (*centripetal acceleration*)

The second form is obtained by eliminating ω from the first form using equation .

ACCELERATION IN SPECIAL RELATIVITY

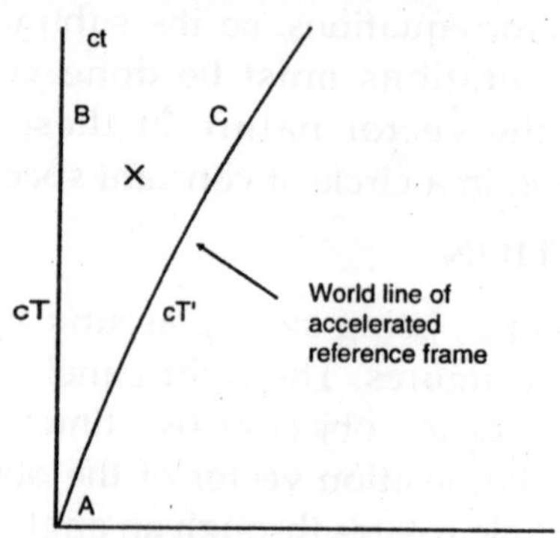

Fig. Acceleration

As noted above, acceleration is just the time rate of change of velocity. We use the above results to determine how acceleration transforms from one reference frame to another. Figure shows the world line of an accelerated reference frame, with a time-varying velocity U(t) relative to the unprimed inertial frame. Defining $\Delta U = U(T) - U(0)$ as the change in the velocity of the accelerated frame (relative to the unprimed frame) between events A and C, we can relate this to the change of velocity, $\Delta U'$, of the accelerated frame relative to an inertial frame moving with the initial velocity, $U(0)$, of the accelerated frame. Applying the equation for the relativistic addition of velocities, we find

$$U(T) = U(0) + \Delta U = \frac{U(0) + \Delta U'}{1 + U(0)\Delta U'/c^2}$$

We now note that the mean acceleration of the reference frame between events A and C in the unprimed reference frame is just $a = \Delta U/T$, whereas the mean acceleration in the primed frame between the same two events is $a' = \Delta U'/T'$. From previous equation

we find that

$$\Delta U = \frac{\Delta U'\left[1 - U(0)^2/c^2\right]}{1 - U(0)\Delta U'/c^2},$$

and the acceleration of the primed reference frame as it appears in the unprimed frame is

$$a = \frac{\Delta U}{T} = \frac{\Delta U'\left[1 - U(0)^2/c^2\right]}{T\left[1 + U(0)\Delta U'/c^2\right]},$$

Since we are interested in the instantaneous rather than the average acceleration, we let T become small. This has three consequences. First, ΔU and $\Delta U'$ become small, which means that the term $U(0)\Delta UU'/c^2$ in the denominator of equation can be ignored compared to 1. This means that

$$a \approx \frac{\Delta U'\left[1 - U(0)^2/c^2\right]}{T}$$

With the approximation becoming perfect as T → 0. Second, the "triangle" with the curved side in figure angled becomes a true triangle, with the result that $T' = T\left[1 - U(0)^2/c^2\right]^{1/2}$. The acceleration of the primed frame with respect to an inertial frame moving at speed $U(0)$ can therefore be written

$$a' = \frac{\Delta U'}{T'} = \frac{\Delta U'}{T\left[1 - U(0)^2/c^2\right]^{1/2}}$$

Third, we can replace $U(0)$ with U, since the velocity of the accelerated frame doesn't change very much over a short time interval.

Dividing both equation results in a relationship between the two accelerations:

$$a = a'\left(1 - U^2/c^2\right)^{3/2},$$

which shows that the apparent acceleration of a rapidly moving object, a, is less than its acceleration relative to a reference frame in which the object is nearly stationary, a′, by the factor $\left(1 - U^2/c^2\right)^{3/2}$. We call this latter acceleration the intrinsic acceleration. This is purely the result of the geometry of spacetime, but it has interesting consequences.

Identifying a with dU/dt, we can integrate the acceleration equation assuming that the intrinsic acceleration a' is constant and that the velocity $U = 0$ at time $t = 0$. We get the following result (verify this by differentiating with respect to time):

$$a't = \frac{U}{\left(1 - U^2/c^2\right)^{1/2}},$$

which may be solved for U/c:

$$\frac{U}{c} = \frac{a't/c}{\left[1 + (a't/c)^2\right]^{1/2}};$$

This is plotted in figure eralier, the velocity would reach the speed of light when $a't/c = 1$. However, the rate at which the velocity increases with time slows as the object moves faster, such that U approaches c asymptotically, but never reaches it.

ACCELERATION, FORCE, AND MASS

We have a good intuitive feel for the concepts of force and mass because they are very much a part of our everyday experience. We think of force as how hard we push on something. Mass is the resistance of an object to acceleration if it is otherwise free to move. Thus, pushing on a bicycle on a smooth, level road causes it to accelerate more readily than pushing on a car. We say that the car has greater mass. We can formalize this relationship with the statement

$$F = ma'$$

where F is the total force on an object, m is its mass, and a is the acceleration resulting from the force.

Three provisos apply to equation. First, it only makes sense in unmodified form when the velocity of the object is much less than the speed of light. For relativistic velocities it is best to write this equation in a slightly different form which we introduce later. Second, the force really must be the total force, including all frictional and other incidental forces which might otherwise be neglected by an uncritical observer. Third, it only works in a reference frame which itself is unaccelerated. We deal below with accelerated reference frames.

ACCELERATED REFERENCE FRAMES

Referring back to the forces being felt by the occupant of a car, it is clear that the forces associated with accelerations are directed opposite the accelerations and proportional to their magnitudes. For instance, when accelerating away from a stoplight, the acceleration is forward and the perceived force is backward. When turning a corner, the acceleration is toward the corner while the perceived force is away from the corner. Such forces are called inertial forces.

The origin of these forces can be understood by determining how acceleration changes when one observes it from a reference frame which is itself accelerated. Suppose that the primed reference frame is accelerating to the right with acceleration A relative to the unprimed frame. The position x' in the primed frame can be related to the position x in the unprimed frame by

$x = x' + X;$

where X is the position of the origin of the primed frame in the unprimed frame. Taking the second time derivative, we see that

$a = a' + A,$

where $a = d^2x/dt^2$ is the acceleration in the unprimed frame and $a' = d^2x'/dt^2$ is the acceleration according to an observer in the primed frame.

We now substitute this into equation and move the term involving A to the left side:

$F - mA = ma'$

This shows that equation is not valid in an accelerated reference frame, because the total force F and the acceleration a' in this frame don't balance as they do in the unaccelerated frame — the additional term $- mA$ messes up this balance.

We can fix this problem by considering $- mA$ to be a type of force, in which case we can include it as a part of the total force . This is the inertial force F which we mentioned above. Thus, to summarize, we can make equation work when objects are observed from accelerated reference frames if we include as part of the total force an inertial force which is equal to $-mA$, A being the acceleration of the reference frame of the observer and the mass of the object being observed.

The inertial force observed in the reference frame of an object moving in circular motion at constant speed. In the case of circular motion the inertial force is called the centrifugal force. It points away from the centre of the circle and just balances the tension in the string. This makes the total force on the object zero in its own reference frame, which is necessary since the object cannot move (or accelerate) in this frame.

General relativity says that gravity is nothing more than an inertial force. This was called the equivalence principle by Einstein. Since the gravitational force on the Earth points downward, it follows that we must be constantly accelerating upward as we stand on the surface of the Earth! The obvious problem with this interpretation of gravity is that we don't appear to be moving away from the centre of the Earth, which would seem to be a natural consequence of such an acceleration. However, relativity has a surprise in store for us here.

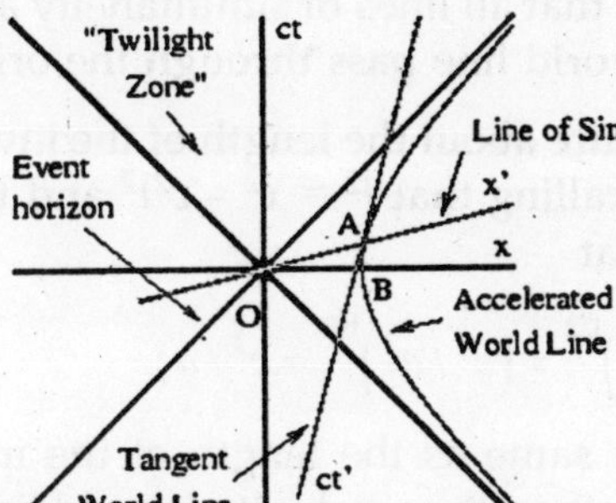

It follows from the above considerations that something can be learned about general relativity by examining the properties of accelerated reference frames. Equation shows that the velocity of an object undergoing constant intrinsic acceleration a (note that we have dropped the "prime" from a) is

$$v = \frac{dx}{dt} = \frac{at}{\left[1+(at/c)^2\right]^{1/2}};$$

where t is the time and c is the speed of light. A function $x(t)$ which satisfies equation (6.21) is

$$x(t) = (c^2/a)\left[1+(at/c)^2\right]^{1/2};$$

The interval OB in figure above is of length $x(0)= c^2/A$.

The slanted line OA is a line of simultaneity associated with the unaccelerated world line tangent to the accelerated world line at point A. This line of simultaneity actually does go through the origin, as is shown in figure. To demonstrate this, multiply equations together and solve for v/c:

$v/c = ct/x$,

ct/x is the slope of the line OA, where (x, ct) are the coordinates of event A. Equation shows that this line is indeed the desired line of

simultaneity, since its slope is the inverse of the slope of the world line, c/v. Since there is nothing special about the event A, we infer that all lines of simultaneity associated with the accelerated world line pass through the origin.

We now inquire about the length of the invariant interval OA in figure. Recalling that $I^2 = x^2 - c^2t^2$ and using equation (6.22), we find that

$$I = \left(x^2 - c^2t^2\right)^{1/2} = \left(c^4 / a^2\right)^{1/2} = c^2 / a;$$

Which is the same as the length of the interval OB. By extension, all events on the accelerated world line are the same invariant interval from the origin. Recalling that the interval along a line of simultaneity is just the distance in the associated reference frame, we reach the astonishing conclusion that even though the object associated with the curved world line of accelerating away from the origin, it always remains the same distance (in its own frame) from the origin. In other words, even though we are accelerating away from the centre of the Earth, the distance to the centre of the Earth remains constant.

GRAVITATIONAL RED SHIFT

Light emitted at a lower level in a gravitational field has its frequency reduced as it travels to a higher level. This phenomenon is called the gravitational red shift. Since experiencing a gravitational force is equivalent to being in an accelerated reference frame, we can use the tools of special relativity to view the process of light emission and absorption from the point of view of the unaccelerated or inertial frame. In this reference frame the observer of the light is accelerating to the right, as indicated by the curved world line in figure, which is equivalent to a gravitational force to the left. The light is emitted at point A with frequency ω by a source which is stationary at this instant. At this instant the observer is also stationary in this frame. However, by the time the light gets to the observer, he or she has a velocity to the right which means that the observer measures a Doppler shifted freqency ω′ for the light. Since the observer is moving away from the source, ω′ < ω, as indicated above.

The relativistic Doppler shift is given by

$$\frac{\omega'}{\omega}=\left(\frac{1-U/c}{1+U/c}\right)^{1/2};$$

so we need to compute *U/c*. The line of simultaneity for the observer at point B goes through the origin, and is thus given by line segment OB in figure of relativity lines early in this chapter. The slope of this line is *U/c*, where *U* is the velocity of the observer at point B. From the figure we see that this slope is also given by the ratio *X/X*. Equating these, eliminating X in favour of $L=(X^2-X'^2)^{1/2}$, which is the actual invariant distance of the observer from the origin, and substituting into equation results in our gravitational red shift formula:

$$\frac{\omega'}{\omega}=\left(\frac{X-X'}{X+X'}\right)^{1/2}=\left(\frac{\left(L^2+X'^2\right)^{1/2}-X'}{\left(L^2+X'^2\right)^{1/2}+X'}\right)^{1/2};$$

If *X′ = 0*, then there is no redshift, because the source is collocated with the observer. On the other hand, if the source is located at the origin, so *X′ = X*, the Doppler shifted frequency is zero. In addition, the light never gets to the observer, since the world line is asymptotic to the light world line passing through the origin. If the source is at a higher level in the gravitational field than the observer, so that *X′ < 0*, then the frequency is shifted to a higher value, i. e., it becomes a "blue shift".

Event Horizons

The 45° diagonal line passing through the origin in figure above is called the event horizon for the accelerated observer in this figure. Notice that light from the "twilight zone" above and to the left of the event horizon cannot reach the accelerated observer. However, the reverse is not true – a light signal emitted to the left by the observer can cross the event horizon into the "twilight zone". The event horizon thus has a peculiar one-way character – it passes signals from right to left, but not from left to right.

SPACE-TIME: CONTINUUMS AND CURVATURE

Relativity views space-time as a continuum; that is, time becomes another dimension of space, making space four-dimensional. This is really common sense-we know that it takes time to move through the physical dimensions of space. (length, width, and heigth) Time is a property of space that is distinct from yet closely related to the other three.

(Althought we know we can move through space in any direction, we can only move forward in time, at least until wormholes are discovered.)

Curvature of space-time can be more difficult to understand. Although the rubber sheet model gives a picture of how it happens, one cannot easily picture the so-called nothingness of space being curved or a non-physical property like time being warped. But relativity (a theory which has yet to be disproved) predicts this curvature and uses gravity as proof of it. In fact, relativity predicts singularities, points in space-time where the curvature reaches infinity. Singularities are the centres of black holes, points of gravity so strong that nothing, not even light, can escape.

According to relativity, the curvature of space-time is gravity, and this raises an intersting paradox: the amount of curvature is governed by the distribution of matter and energy in the univers, but this distribution is determined by the curvature of space-time. Thinking back to the rubber sheet model, this becomes an obvious truth. A large object, such as a shot put, would cause a large indentation that would cause a marble to roll in. Or, if an area of rubber sheet space contained only small objects, several marbles might roll together to form an area of 'strong gravity', which would in turn attract more objects.

Besides creating the all-important force of gravity, space-time curvature also makes a major dent (no pun intended) in one of our most common beliefs from geometry and family vacations: in curved space-time, a straight line is not always the shortest distance between two points. But that's a different topic.

In order to move through the physical dimensions of space, one must also move through time. In this graph, time surrounds the axes, and although the lines could be extended so that we could move in any physical direction, we do not extend them because we do not know how to move backwards in time.

Parallel universes would seem to be the most unlikely prediction of relativity, but their proposal is just as fact-based as those dealing with black holes.

The idea of parallel universes was taken seriously after experiments with properties of light. Night exhibits wave-particle duality-it acts like both a particle and a wave. But even this odd characteristic did not account for the surprising results scientists noted in 'double siit tests'. A beam of light would be shined at a screen through another screen with two slits in it. When both slits were open, it seemed as if an electron would pass through both at the same time; and with one open, some areas of the screen would get more light then when both were open. While some of this could be attributed to particle-wave duality, in some cases it seemed almost as though the possibilities of what the light could do was affecting the outcome.

The only way for this to happen is for all the possiblities to exist at once. This idea is the backbone of the parallel universe theory-that all the possibilities for every action exist as an alternate universe. At the moment when an outcome is needed, they can overlap, or one of the separate universes will bend its solution.

Because there are an infinite number of possibilities, there is an infinite number of parallel universes, and therefore, an infinite number of you. A parallel universe for you oversleeping this morning, losing your homework, wearing different clothes... it goes on. For every interaction, there is a myriad of possibilites.

This all seems very far-fetched from Einstein's theory, but Einstein actually speculated and believed in parallel universes. He helped to discover the Einstein-Rosem Bridge-the area at the singularity of a rotating black hole where one can cross into a parallel universe.

Which is really the heart of the problem-reaching the singularity of a rotating black hole. But you'll have to read about black holes to find out why.

DEFINITION OF GRAVITY

If gravity can be defined as the curvature of space rather than a force of attraction, why does not a bullet shot out of a gun, say perpendicular to the Earth's crust, and a ball thrown by me on the same trajectory (but obviously at a much slower speed) follow the same curve?, If gravity is actually curved space and if falling objects are simply following the natural curves of space why does each object have its own curve?

This is a great question which goes to the heart of why Einstein said gravity is the curvature of space-time, rather than just the curvature of space.

Consider the diagram below, which shows the situation you describe in your question. The ball and the bullet both start off along the same path in space (i.e. they both start off moving horizontally away from the person's head). However, as we know, their paths quickly diverge — the bullet will travel much farther before hitting the ground than the ball does.

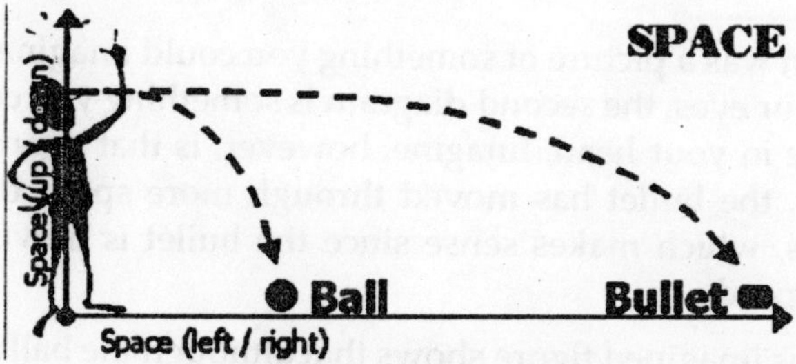

Fig. Gravity Experiment

This might not seem strange at first, but under Einstein's reconception of gravity, it's a major problem! Einstein's idea was that there is no such thing as a "force" of gravity which pulls things to the Earth; rather, the curved paths that falling objects appear to take are an illusion brought on by our inability to perceive the underlying curvature of the space we live in. The objects themselves are just moving in straight lines.

If this is true, however, then the ball and the bullet which start off on the same path should logically continue on the same path. After all, if you imagine walking on a curved surface such as the Earth, if you start off walking in a straight line towards the east and your friend starts from the same location running in a straight line towards the east, you will both follow the exact same path! It doesn't matter how fast you're going; you will both (eventually) reach the same location. So why don't the ball and the bullet wind up in the same location too?

The only way to get around this problem in Einstein's theory is to say that it is not just "space" which is curved; rather, it is "space-time." To understand this, it's helpful to look at the illustration below. This illustration shows the same ball and bullet as above, only now they are plotted on a diagram of space and time together ("space-time").

The horizontal axis is the same as it was before; it represents distance in the left and right direction. The vertical axis, however, no longer represents distance in the up and down direction like it did before; instead, the vertical axis represents time, or, specifically, how much time has elapsed since the ball and bullet were released. Whereas the first diagram was a picture of something you could imagine seeing with your eyes, the second diagram is something you can only imagine in your head. Imagine, however, is that at any point in time, the bullet has moved through more space than the ball has, which makes sense since the bullet is moving at a faster speed.

This imagined figure shows that although the ball and the bullet start off along the same direction in space, they actually start off along different directions in space-time. So if we agree that space-time, and not space, is the proper arena in which to consider the question, then we can understand why the ball and the bullet don't wind up in the same place at the end of their trip. Just like it's not surprising that if you start off walking east and your friend starts off walking northeast, you will end up in different places, it also isn't surprising that the ball and the bullet end up in different places, since they started off in different directions!

Furthermore, let us consider what happens if two objects start off along the same path in space-time. You should be able to convince yourself that the only way this can happen is if the two objects start off in the same direction and at the same speed. For example, we might fire two different bullets from the same gun, with each bullet made of a different material. Or we might fire a bullet and a cannonball from the same point in space, with each starting out at the same speed. It turns out that in these cases, if no other forces such as air resistance act on the objects being fired, they will follow the exact same paths and hit the ground at the exact same time and exact same place. This is completely consistent with Einstein's theory – the path that an object takes through space-time doesn't depend on the mass of the object or on the material it is made of; it only depends on the initial direction that the object starts off in.

In some sense, therefore, what Einstein's theory tells us is that we really need to consider space-time, rather than space, as the fundamental "playing field" upon which the events of the universe occur. It is a profound realization to understand this fact – all the objects around us actually exist in a realm of "space-time" that is much more complex than the simple realm of space in which we perceive them.

THE SPEED OF GRAVITY

You must have been thinking about gravity and would like to ask if we actually recorded the speed in which it moves.

This is a very interesting and timely question. There was recently an experiment which aimed to measure the speed of gravity, and there has been some disagreement among scientists over the interpretation of the results.

In the theory of relativity, the speed of gravity should be equal to the speed of light, since the theoretical "particles" that carry gravity (sometimes called gravitons) are massless particles, just like photons (the particles that carry light). The light from the Sun takes 8 minutes to reach the Earth, so that if the Sun suddenly disappeared it would take 8 minutes before it got dark. Similarly the Earth would also feel the effects of the Sun's gravity for 8 minutes after it magically vanished.

In September 2002, two US scientists made some very accurate measurements of the position of a quasar as it passed behind Jupiter. They argued that the exact amount of apparent motion of the quasar (as the path of the radio waves from it was bent in Jupiter's gravitational field) depended on both the speed of light and the speed of gravity. The measurements they took then proved that the speed of gravity is the same as that of light; ruling out some of the more bizarre modifications to the laws of gravity which have been proposed, and further backing General Relativity.

However, other astronomers disagree that the experiment is able to measure the speed of gravity, arguing that the effect is much smaller than the scientists claim and that (in effect) they got their arithmatic wrong when they decided that the speed of gravity did come into the equations. They are not claiming that the speed of gravity is different to that of light, just that it could not be measured in the experiment.

The speed of gravity should be equal to the speed of light, and that there is a ongoing disagreement over whether or not that has actually been measured.

Have gravitational waves been proven to exist 100%? Or are astrophysicists only about 99.9% sure that they are real.

Gravitational waves are predicted by General Relativity, which is one of the most supported theories in science. Nevertheless, they have never actually been detected, mostly because of the difficulty involved in the endeavor. Astronomers do watch the orbits of close binary pulsars, which are predicted to emit gravitational waves, and find that the orbits decay at exactly the correct rate predicted by the theory of gravitational radiation and have no other explanation available. So that is the best evidence that gravitational waves exist, and is probably what you were reading about. It's pretty good evidence.

According to Weisstein's world of Physics, a graviton is "a theoretical particle having no mass and no charge that carries the gravitational force."

In quantum mechanics, we know that light comes to us in massless packets, or particles, called photons, which carries electromagnetic force with it. Because this worked so well and is essential to the theory of quantum mechanics, people hypothesized the existence of a similar particle which carries the gravitational force.

MATTERS OF GRAVITY

Apart from rare and generally regrettable moments of free-fall, we spend our entire lives under the influence of the Earth's gravity, yet rarely, if ever, do we experience the universal nature of gravitation. It's a tremendous philosophical leap from "stuff falls" to "everything in the universe attracts everything else". That leap, made by Isaac Newton in the 17th century, not only allowed understanding the motion of the Moon and the planets, but inoculated in Western culture the idea that the universe as a whole was governed by laws humans could discover. This realisation fueled the

Enlightenment and the subsequent development of science and technology.

This page presents a "basement science" experiment which reveals the universality of gravitation by demonstrating the gravitational attraction between palpable objects on the human scale. The experiment deliberately uses only the crudest and most commonplace materials, permitting anybody who's so inclined to perform it. Einstein's 1915 theory of General Relativity explains gravitation as spacetime curvature created by matter and energy. So, by demonstrating how every object in the universe attracts everything else, we're bending spacetime in the basement.

But, if gravitation is ubiquitous, why was it not discovered millennia before Newton's 1687 Philosophiæ naturalis principia mathematica? The reason lies in the extraordinary weakness of the gravitational force.

Feeble Attraction

Now you might say, "What do you mean, a weak gravity! You fell down a flight of stairs a couple of years ago, and gravity sure did not feel weak to you!". And yet, of the four forces of nature known to physics, gravitation is the weakest, by the mindboggling factor of 4.17×10^{42} (4 followed by 42 zeroes) times weaker than the electromagnetic force.

The stark difference in the strength of the electromagnetic and gravitational forces is evident in the picture to the left. The bright square in the jaws of the pliers is a 4 mm cubical magnet. It is lifting a spherical steel pétanque ball which weighs 550 grams. we're pitting our valiant little magnet, with a volume of 0.064 cubic centimetres, weighing less than one gram, pulling up with the electromagnetic force, against the entire Earth, pulling down with gravity. And the winner is...the magnet. A one gram magnet out-pulls the Earth, which weighs 5.9736×10^{27} grams and has a volume of more than 10^{27} cubic centimetres.

(The apparent discrepancy between the ratio of masses of the Earth and the magnet and the 4.17×10^{42} strength ratio of electromagnetism and gravitation is due to the fact that only an infinitesimal fraction of the mass of the magnet contributes to the [electro]magnetic attraction on the ball, while every gram of the Earth's mass exerts gravitational force. To obtain the correct ratio of force strengths, one must compare the gravitational attraction between two electrons at a given distance with the electromagnetic repulsion resulting from their charge. This calculation arrives at the correct strength ratio for the two forces.)

If gravity were not so weak compared to the electromagnetic force, you wouldn't be reading this page; it's only because the electromagnetic force that bonds the atoms in your body together so easily defeats the Earth's gravity that you, along with all other solid objects, don't slump into a puddle and eventually merge into a perfectly spherical (actually, slightly ellipsoidal thanks to rotation) planet.

If your browser supports JavaScript, you can use the following calculator to determine the gravitational force between any two objects. The gravitational force between two masses m and m' whose centres of gravity are separated by a distance r is given by:

$$F = G\frac{mm'}{r^2}$$

Since most folks don't encounter forces in dynes on an everyday basis, not to mention small fractions of a dyne, it's useful to express this force in terms of the weight of something one can visualise. We can calculate the mass of an object from its weight at the Earth's surface merely by solving the gravitational force equation above for m and plugging in the Earth's mass, $m_e = 5.9736\times10^{27}$ grams, and the Earth's equatorial radius, $r_e = 6.37814\times10^{8}$ cm:

$$m = \frac{r^2F}{Gm_e}$$

The value below, then, is the mass of an object whose weight at the Earth's surface equals the gravitational force displayed in the box above. This value is automatically updated when you press the "Calculate" button.

Even though the force of gravity between objects of modest mass is palpable compared to the weight of objects one can see, detecting such a tiny force seems a daunting, if not hopeless, endeavour for the basement tinkerer. Certainly, painstakingly designed and constructed laboratory apparatus has allowed measuring the gravitational constant to great precision and verifying the equivalence of gravitational and inertial mass, and precision gravitometers are routinely used in oil and gas exploration and mineral prospecting, but we're trying to see if we can experience the universal attraction of gravity without any high tech, high budget gear.

Measuring tiny gravitational forces would be easy if we were in deep space, far from any massive bodies. The only forces on objects in our space laboratory, then, would be those entirely under our control. As long as we made sure none of the objects we were experimenting with were magnetic or electrically charged (easily arranged, assuming they are conductive, simply by bringing them into contact so all excess charges equilibrate), the only force remaining between objects would be gravity, so however weak it be, we need only be sufficiently patient to observe its effects. (The other two forces, the strong and weak nuclear interactions, are limited in range to distances on the order of the size of an atomic nucleus and can be neglected on the human scale.)

What we would like to do, then, is cancel the Earth's gravity so that the much smaller gravitational forces between objects that fit in the basement become evident.

Differential Cleverness

One of the great all-purpose sledgehammers in the toolbox of physicists and engineers is differential measurement; in other words, don't worry about the absolute value of something, but only the difference between things you can measure. For example, it is common practice for linemen repairing high-voltage power transmission lines to work on them, without cutting power, from insulated baskets raised by a crane. As long as the lineman is insulated from the ground, only the voltage difference between his hands and the line he's working on matters; after attaching the basket to the line, this is zero, so he might as well be repairing a grounded conductor. Now if, while working on a conductor at, say, 200,000 volts above Earth potential, he should happen to touch the tower, grounded to Earth, that would make for a really bad day. The trick is keeping the difference small; you can live your entire life at 1 million volts, and as long as everything around you is near that value, there is no way, even in principle, you could discover the absolute potential. This is the consequence of all the forces of physics being gauge invariant: absolute values don't exist — only differences matter.

The Torsion Balance

What we were looking then, then, is a device which responds only to differences in gravitational attraction, canceling out the much stronger constant gravitational attraction of the Earth. We need look no further than a slightly modified version of the same device Henry Cavendish used in 1798 to first measure the gravitational constant, G in the equations above. Ever since, the torsion balance has been the primary tool used both for measuring the gravitational constant and testing the equivalence principle, which states that all bodies experience the same gravitational force regardless of composition; Einstein's General Relativity showed this to be a fundamental consequence of the structure of space and time.

State of the art torsion balances have measured the gravitational constant to better than one part per million and confirmed the equivalence principle to more than 11 decimal places. This requires extraordinarily refined and delicate laboratory apparatus and experimental design, in which a multitude of subtle effects must be compensated for or canceled out. We, however, aren't going to measure anything — we were only interested in observing universal gravitation. This allows simplifying the torsion balance to something we can set up in the basement.

The principle of the torsion balance is extremely simple. Suspend a horizontal balance arm from a vertical elastic fibre. At each end of the balance arm are masses, much denser than material of the arm, which respond to the gravitational force. Once the suspending fibre, balance arm, and weights are set up and brought into balance, the downward force of gravitation acts equally on every component. The balance arm is then free to rotate without any hindrance from the Earth's gravity. It is constrained only by air friction and the torsional strength of the support fibre – its resistance to being twisted. We can then place test masses near the ends of the balance arm and observe whether the gravitational attraction between them and the masses on the arm causes the balance arm to move. When measuring the gravitational constant one must precisely calibrate the torsional strength of the fibre, but to simply observe gravitation we need only make sure the fibre is sufficiently limp to allow the gravitational force to overcome its resistance to twisting.

In practice, the balance arm is so free to move that once any force sets it into motion, it oscillates for a long period, spinning round and round if free or bouncing back and forth off the stops if constrained. To avoid this we need to damp the system so kinetic energy acquired by the bar is more rapidly dissipated. Well, nothing's more damp than water, so we add a water brake to the arm which turns in a fixed reservoir. The resulting drag as the balance arm moves is much greater than air resistance and frictional losses in the fibre, and reduces the oscillations to a tolerable degree.

Mutual Attraction of Tide

What we have demonstrated by these experiments is the universality of gravitation; there is nothing special about the Earth that makes objects fall toward it. Everything attracts everything else; the Earth's attraction is greater simply because the Earth is so much more massive than the objects we encounter in everyday life. Only by canceling out the Earth's gravitation by means of a torsion balance were we able to observe the gravitational attraction between masses of less than a kilogram.

The universality of gravitation means that every object in the universe is interlinked in a web of mutual attraction; the universe is transparent to gravitation. The most distant galaxies exert a pull on you, as you do upon them – immeasurably tiny to be sure, but present just the same. From a practical standpoint, universality means there's no way to shield your torsion balance from the gravitational attraction of masses in its vicinity; you can only set it up sufficiently far from other massive objects so the attraction of the test masses predominates. One interesting massive object to consider is yourself .

Using the gravitational force calculator, we can compute the gravitational attraction between the 338 gram mass at the end of the balance beam and the 740 gram test mass at the 14 cm distance when the beam is at the midpoint between the masses to be 0.000085 dynes. Now suppose you're crouching down in order to move the test masses, with your centre of gravity one metre from the closer test mass, and that you weigh 65 kg. Plugging these numbers into the calculator shows that your own gravitational attraction on the nearer end of the beam is 0.000147 dynes, 1.7 times as great as that of the test mass. Your actual influence on the motion of the balance arm is less, however, since what matters is the difference in force exerted on the masses at the two ends of the balance arm. Since your centre of gravity is more distant than the test masses, the difference is less.

Assume the centres of gravity of the two masses on the balance arm are 25 cm apart, and that you're crouching so the arm makes a 45° angle with your centre of gravity, one metre from the centre of the arm. The nearer mass is then 17.68 cm closer than the more distant one and the difference in gravitational attraction (or tidal force) on the two masses is the difference in attraction on a mass 91.16 cm distant and one 108.84 cm away. The calculator gives the attraction on the near end of the arm as 0.0001764 dynes and the far end as 0.0001238 dyne, with a difference of 0.0000527 dynes. Now recall that the force exerted by the test mass was 0.000085 dynes, only 1.6 times as large, so even taking into account the reduced tidal influence due to your greater distance, the force you exert on the balance cannot be neglected. This makes it essential to remotely monitor the experiment so your own mass doesn't disrupt it.

In practice, air currents due to your motion and resulting from convection driven by your body's temperature being above room temperature may exert greater forces on the balance arm than the gravitational field generated by your mass.

An Historical Speculation

Nineteen centuries elapsed between the death of Archimedes in 212 B.C. and the publication of Newton's Principia in 1687. Given the philosophical implications of Newton's theory, it's interesting to speculate what might have happened had Archimedes discovered the universal nature of gravitation.

To do this, he would have had to suspect that attraction was universal, suggest an experiment to confirm this, and perform that experiment, with results validating the hypothesis. Here is information in Archimedes' possession which might have suggested the universality of gravitation.

- The Earth is a sphere. The shape of its shadow on the Moon during a lunar eclipse demonstrated this, and was confirmed by the next item. The assertion that "the ancients thought the world was flat" is nonsense — Columbus didn't "discover the world was round": he discovered that his own estimate of the diameter of the world was wrong by a factor of two compared to that available to Archimedes; if he hadn't inadvertently discovered the New World, he and his unfortunate crew would have died of starvation far from the coast of China.

- The approximate radius of the Earth. Around 250 B.C., by measuring the difference in the angle of sunlight at noon on the June Solstice, which illuminated the bottom of a well at Syene (now Aswan) Egypt near the Tropic of Cancer, with the length of the shadow cast on the same date and time by a vertical pillar in Alexandria, a known distance to the North, Eratosthenes determined the Earth's circumference. Archimedes corresponded extensively with Eratosthenes and other scholars in Alexandria, and knew of this result. Archimedes himself calculated the value of Pi as between 3 10/71 and 3 1/7, with a mean value of 3.14185, allowing accurate computation of the Earth's radius from the circumference.
- The approximate mass of the Earth. Assuming the Earth to have the same density as common rocks such as limestone (2.7 g/cm^3) gives an estimate within a factor of two of the correct value. The actual density of the Earth is 5.52 g/cm^3.
- How to calculate with very large numbers. In The Sand Reckoner ($\psi\alpha\mu\mu\iota\tau\eta\varsigma$) in 215 B.C., Archimedes invented a positional number system which allowed writing and calculating with arbitrarily large quantities, which he demonstrated by calculating not only how many grains of sand would fill the volume of the Earth, but how many grains of sand would fill the entire universe (which the Greeks estimated to be about one light year in diameter). The latter number, about 10^{63}, is comfortably larger than any of the quantities associated with gravitation.
- The existence of electrostatic and magnetic forces which appeared to act at a distance. The inverse square behaviour of these forces was not known in antiquity, however.

Suppose then that, given these facts, Archimedes embarked upon the following chain of reasoning.

- Objects fall, not in a fixed direction, but toward the centre of our world. If they fell in a fixed direction, if I dropped a rock down a well in the south of Egypt and a well in Syracuse, separated by a substantial percentage of the world's circumference, one would hit the wall of the well before striking the bottom. This doesn't happen, so objects fall everywhere toward the centre of the world.
- Why does the world attract falling objects? Is there something special which endows it with this property? Yet the world seems to be made of the same substances as everything else. What is the difference between the world and a rock?
- The world is much larger than a rock! Perhaps every object attracts every other. We only feel the world's attraction because it is so large.
- But if this is so, might the celestial bodies be objects no different from the world and its inhabitants, and subject to the same forces?
- If attraction is universal, might an artificer be able to build a device to show it?
- Such a device must be isolated from falling down. Perhaps a horizontal balance, free to turn in either direction, with weights at each end to be attracted to objects in their vicinity....

The Archimedes Apparatus

It seems plausible, then, given the knowledge at hand and a chain of inference which, in retrospect at least, appears straightforward, that Archimedes could have suspected the universality of gravitation. But could he have demonstrated it? Unlike many scholars in ancient Greece who contented themselves with philosophical arguments, Archimedes was an intensely practical man, renowned as a military engineer as well as a mathematician and philosopher. His laws of the lever and buoyancy were tested experimentally, and so we should expect he would subject any inference about gravitation to experimental confirmation. Now that we've succeeded in bending spacetime in the basement with common household materials of the late 20th century, let's see if the experiment can be done using only materials Archimedes might have employed.

Gravity as Curved Space

Newton's Law of Universal Gravitation had three basic flaws:

1. It gave the wrong prediction for the precession of the perihelion of Mercury's orbit. Mercury's orbit is elliptical, as predicted by Newton's theory of gravity, but the ellipse doesn't stay in precisely the same place all the time. It precesses, which is to say that as Mercury orbits the sun, the entire ellipse rotates about the focal point (i.e. the sun) as shown in the Figure below. This precession is very small; only 570 seconds of arc per century. A second of arc is 1/360 of a degree. Most of this precession could be understood in the context of Newton's theory of gravity by taking into account perturbations of the orbit due to the presence of other planets. However, once this was done, there still remained a discrepancy of about 40 seconds of arc per century between the prediction, and the observed value. This discrepancy was a complete mystery to scientists at the turn of the century. They even went as far as postulating the existence of an unseen planet (Vulcan) on the far side of the Sun in order to explain it. It was not until Einstein published his work on the general theory of relativity that the perihelion shift of Mercury was truly understood.
2. It did not explain why the gravitational force on an object was proportional to its inertial mass. In other words it did not explain why gravitational acceleration is independent of the mass or composition of an object.

3. It was inconsistent with the Special Theory of Relativity. That is, if an instantaneous force of gravitational attraction exists between two objects then information about the location of one object would be transmitted to another object instantaneously by changes in the gravitational force. Thus it would be possible to send information faster than the speed of light.

The heavier the star, the deeper the gravitational well, and the more curved space becomes.

Einstein's General Theory of Relativity did away with all three of the above problems, and at the same time it radically altered physicists' view of the Universe. The main features of General Relativity are:

1. Space and space-time are not rigid arenas in which events take place. They have form and structure which are influenced by the matter and energy content of the universe.
2. Matter and energy tell space (and space-time) how to curve.
3. Space tells matter how to move. In particular small objects travel along the straightest possible lines in curved space (space-time). (Note the above descriptions of General Relativity are due to John Wheeler.)

In curved space the rules of Euclidean geometry are changed. Parallel lines can meet, and the sum of the angles in a triangle can be more, or less than 180 degrees, depending on how space is curved. Einstein's theory gave a correct prediction for the perihelion shift of Mercury. It also explained why objects fall independent of their mass: they all follow the same straightest possible line in curved space-time. Finally, in Einstein's theory the instantaneous gravitational force is replaced by the curvature of spacetime. Moving a mass causes ripples to form in this curvature, and these ripples travel with the same speed as light. Thus, a distant mass would not feel any instantaneous change in the gravitational force, and special relativity is not violated.

Chapter 5

Inertial Frames

SPACE AND TIME: INERTIAL FRAMES

A "frame of reference" is a standard relative to which motion and rest may be measured; any set of points or objects that are at rest relative to one another enables us, in principle, to describe the relative motions of bodies. A frame of reference is therefore a purely kinematical device, for the geometrical description of motion without regard to the masses or forces involved. A dynamical account of motion leads to the idea of an "inertial frame," or a reference frame relative to which motions have distinguished dynamical properties. For that reason an inertial frame has to be understood as a spatial reference frame together with some means of measuring time, so that uniform motions can be distinguished from accelerated motions. The laws of Newtonian dynamics provide a simple definition: an inertial frame is a reference-frame with a time-scale, relative to which the motion of a body not subject to forces is always rectilinear and uniform, accelerations are always proportional to and in the direction of applied forces, and applied forces are always met with equal and opposite reactions. It follows that, in an inertial frame, the centre of mass of a system of bodies is always at rest or in uniform motion. It also follows that any other frame of reference moving uniformly relative to an inertial frame is also an inertial frame. For example, in Newtonian celestial mechanics, taking the "fixed stars" as a frame of reference, we can determine an (approximately) inertial frame whose centre is the centre of mass of the solar system; relative to this frame, every acceleration of every planet can be accounted for (approximately) as a gravitational interaction with some other planet in accord with Newton's laws of motion.

This appears to be a simple and straightforward concept. By inquiring more narrowly into its origins and meaning, however, we begin to understand why it has been an ongoing subject of philosophical concern. It originated in a profound philosophical consideration of the principles of relativity and invariance in the context of Newtonian mechanics. Further reflections on it, in different theoretical contexts, had extraordinary consequences for 20th-century theories of space and time.

simple and straightforward concept. By inquiring more narrowly into its origins and meaning, however, we begin to understand why it has been an ongoing subject of philosophical concern. It originated in a profound philosophical consideration of the principles of relativity and invariance in the context of Newtonian mechanics. Further reflections on it, in different theoretical contexts, had extraordinary consequences for 20th-century theories of space and time.

RELATIVITY AND REFERENCE FRAMES IN CLASSICAL MECHANICS

THE ORIGINS OF GALILEAN RELATIVITY

The term "reference frame" was coined in the 19th century, but it has a long prehistory, beginning, perhaps, with the emergence of the Copernican theory. The significant point was not the replacement of the earth by the sun as the centre of all motion in the universe, but the recognition of both the earth and the sun as merely possible points of view from which the motions of the celestial bodies may be described. This implied that the basic task of Ptolemaic astronomy — to represent the planetary motions by combinations of circular motions — could take any point to be fixed, and that, as Copernicus suggested in the opening arguments of "On the revolutions of the heavenly spheres," the choice of any particular point required some justification on other than astronomical grounds. As the basic programme of Ptolemy and Copernicus gave way to that of early classical mechanics, this equivalence of points of view was made more precise and explicit. Galileo demonstrated that the Copernican view does not contradict our experience of a seemingly stable earth, through a principle that, in the precise form that it takes in Newtonian mechanics, has become known as the "principle of Galilean relativity": mechanical experiments will have the same results in a system in uniform motion that they have in a system at rest. Therefore the experiments claimed as evidence against Copernicus – e.g., that a stone dropped from a tower falls to the base of the tower, instead of being left behind – would happen just as they do whether the earth were moving or not, provided that the motion is sufficiently uniform.

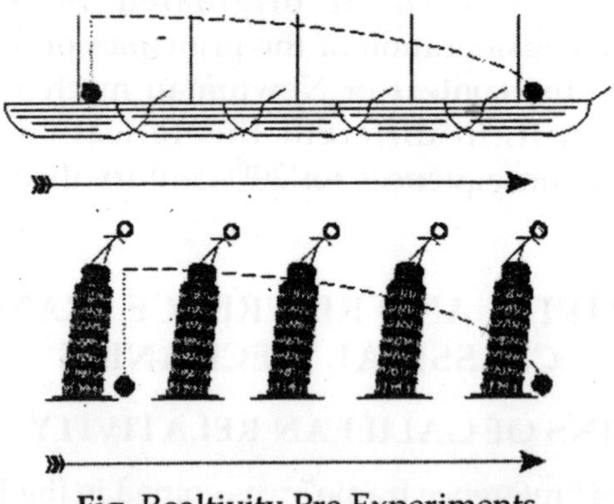

Fig. Realtivity Pre Experiments

Galileo's Argument

If the earth is rotating sufficiently uniformly, a stone dropped from the tower will fall straight to the base, just as a stone dropped from the mast of a uniformly moving ship will fall to the foot of the mast. In both cases the stone's vertical motion will be smoothly composed with its horizontal motion. Hence a sufficiently uniform motion will be indistinguishable from rest.

PHILOSOPHICAL CONTROVERSY OVER ABSOLUTE AND RELATIVE MOTION

Leibniz, later, articulated a more general "equipollence of hypotheses": in any system of interacting bodies, any hypothesis that any particular body is at rest is equivalent to any other. Therefore neither Copernicus' nor Ptolemy's view can be true – though one may be judged simpler than the other – because both are merely possible hypothetical interpretations of the same relative motions. This principle clearly defines (what we would call) a set of reference frames, differing in their arbitrary choices of a resting point or origin, but agreeing on the relative positions of bodies at any moment and their changing relative distances through time.

For Leibniz and many others, this general equivalence was a matter of philosophical principle, founded in the metaphysical conviction that space itself is nothing more than an abstraction from the geometrical relations among bodies. In some form or other it was a widely shared tenet of the 17th-century "mechanical philosophy". Yet it was flatly incompatible with physics as Leibniz himself, and the other "mechanists," actually conceived it. For the basic programme of mechanical explanation depended essentially on the concept of a privileged state of motion, as expressed in the assumption that bodies maintain a state of rectilinear motion until acted upon by an external cause. Thus their fundamental conception of force, as the power of a body to change the state of another, likewise depended on this notion of a privileged state. This dependence was clearly exhibited in the vortex theory of planetary motion, in which every orbit was explained by the balance between the planet's inherent centrifugal tendency (its tendency to follow the tangent to the orbit) and the pressure of the surrounding medium.

For this reason, the notion of a dispute between "relativists" or "relationists" and "absolutists" or "substantivalists", in the 17th century, is a drastic oversimplification. Newton, in his controversial Scholium on space, time, and motion, was not merely asserting that motion is absolute in the face of the mechanists' relativist view; he was arguing that a conception of absolute motion was already implicit in the views of his opponents — that it was implicit in their conception, which he largely shared, of physical cause and effect. The general equivalence of reference-frames was implicitly denied by a physics that understood forces as powers to change the states of motion of bodies.

Newton therefore held that physics required the conception of absolute space, a distinguished frame of reference relative to which bodies could be said to be truly moving or truly at rest. Assuming, as both Newton and Leibniz did, that states of motion could be distinguished by their causes and effects, the distinguished status of this frame of reference is physically well founded — and metaphysically well-founded for a metaphysics that, like Newton's or Leibniz's, takes force to be a well-founded notion. On Leibniz's conception of force, in particular, a given force is required to generate or to maintain a given velocity — for objects "passively" resist motion, but maintain their states of motion only by "active" force — so that, on dynamical grounds, "every body truly does have a certain amount of motion, or, if you will, force." This implies that there is in principle a distinguished frame of reference in which the velocities of bodies correspond to their true velocities, i.e. to the amounts of moving force that they truly possess, and it implies that in any frame that is in motion relative to this one, bodies will not have their true velocities. In short, such a conception of force, if it could be applied physically, would give a precise physical application of Newton's conception of absolute space.

GALILEAN RELATIVITY IN NEWTONIAN PHYSICS

The difficulty with Newton's view of absolute space comes from the Newtonian conception of force. If force is defined and measured solely by the power to accelerate a body, then obviously the effects of forces — in short, the causal interactions within a system of bodies — will be independent of the velocity of the system in which they are measured. So the existence of a set of equivalent "inertial frames" is imposed from the start by Newton's laws. Suppose that we determine for the bodies in a given frame of reference – say, the rest frame of the fixed stars – that all observable accelerations are proportional to forces impressed by bodies within the system, by equal and opposite actions and reactions among those bodies. Then we know that these physical interactions will be the same in any frame of reference that is in uniform rectilinear motion relative to the first one. Therefore no Newtonian experiment will be able to determine the velocity of a body, or system of bodies, relative to absolute space. In other words, there is no way to distinguish absolute space itself from any frame of reference that is in uniform motion relative to it. Newton thought that a coherent account of force and motion requires a background space consisting of "places" that "all keep given positions in relation to one another from infinity to infinity". But the laws of motion enable us to determine an infinity of such spaces, all in uniform rectilinear motion relative to each other, and furnish no way of singling out any one as "immovable space."

Oddly enough, no one in the 17th century, or even before the late 19th century, expressed this equivalence of reference-frames more clearly than Newton himself. Newton explicitly derived it from the laws of motion as Corollary V:

Corollary V

When bodies are enclosed in a given space, their motions in relation to one another are the same whether the space is at rest or whether it is moving uniformly straight forward without circular motion.

This is the first clear statement of the Galilean relativity principle. It implied that the dispute between the heliocentric and geocentric views of the universe was mistakenly framed: the proper question about "the system of the world" was not "which body is at rest in the centre?" but "where is the centre of gravity of the system, and which body is closest to it?" For in a system of orbiting bodies, only their common centre of gravity will be unaccelerated, and by Corollary V, the motions of the bodies in the system will be the same, whether its centre of gravity is at rest or in uniform rectilinear motion. The system is indeed approximately Keplerian, since the sun has by far the greatest mass and is therefore little disturbed from the centre of gravity, which is therefore very close to the common focus of the approximately Keplerian ellipses in which the planets orbit the sun. But by Corollary V, the nearly-Keplerian structure of the system is completely independent of the system's state of motion in absolute space.

The Galilean relativity principle thus expressed the insight that different states of uniform motion, or different uniformly-moving frames of reference, determine only different points of view on the same physically objective quantities, namely force, mass, and acceleration. We can see this insight expressed more explicitly in Newton's understanding of inertia. For Leibniz (among others), as we saw, moving force, the power of a body to change the motion of another, was determined by velocity. It was therefore seen as an active power, fundamentally different from the passive power of a resting body to resist any change of position. Newton, in contrast, understood the "force of inertia" as a Galilei-invariant quantity:

> A body exerts this force only during a change of its state, caused by another force impressed upon it, and the exercise of this force is, depending on viewpoint, both resistance and impetus: resistance in so far as the body, in order to maintain its state, strives against the impressed force, and impetus in so far as the same body, yielding only with difficulty to the force of a resisting obstacle, endeavors to change the state of that obstacle. Resistance is commonly attributed to resting bodies and impetus to moving bodies; but motion and rest, in the popular sense of the term, are distinguished from each other only by point of view, and bodies commonly regarded as being at rest are not always truly at rest.

Newton thus recognized the powers distinguished by Leibniz as the same thing seen from different points of view.

THE LINGERING PROBLEM OF ABSOLUTE SPACE

Newton understood the Galilean principle of relativity with a degree of depth and clarity that eluded most of his "relativist" contemporaries. It may seem bizarre, therefore, that the notion of inertial frame did not emerge until more than a century and a half after his death. He had identified a distinguished class of dynamically equivalent "relative spaces," in any of which true forces and masses, accelerations and rotations, would have the same objectively measured values. Yet these spaces, though empirically indistinguishable, were not equivalent in principle; evidently Newton conceived them as moving with various velocities in absolute space, though those velocities could not be known. Why should not he, or someone, have recognized the equivalence of these spaces immediately?

This is not the place for an adequate answer to this question, if indeed one is possible. For much of the 20th century, the accepted answer was that of Ernst Mach: Newton lived in an age "deficient in epistemological critique," and so was unable to draw the conclusion that these empirically indistinguishable spaces must be equivalent in every meaningful sense, so that no one of them deserves even in principle to be designated as "absolute space." Yet even those whom the 20th century credited with more sophisticated epistemological views, such as Leibniz, evidently had difficulties understanding force and inertia in a Galilei-invariant way, despite a philosophical commitment to relativity. Perhaps it suffices to say that to abandon the intuitive association of force or motion with velocity in space, and to accept an equivalence-class structure as the fundamental spatiotemporal framework, requires a level of abstraction that became possible only with the extraordinary development of mathematics, especially of a more abstract view of geometry, that took place in the 19th century. In the 17th century only Christiaan Huygens came close to expressing such a view; he held that not velocity, but velocity-difference, was the fundamental dynamical quantity. He therefore understood, for example, that the "absoluteness" of rotation had nothing to do with velocity relative to absolute space, but arose from the difference of velocity among different parts of a rotating body – a difference which would, evidently, be the same irrespective of the velocity of the body as a whole in absolute space. But of this Huygens gave only the merest suggestion, in manuscripts that remained unpublished for two centuries. The concept of inertial frame therefore emerged only in the late 19th century, when, as we shall see, it did not seem to be of any great immediate importance.

19TH-CENTURY ANALYSES OF THE LAW OF INERTIA

The development of this concept began with a renewed critical analysis of the notion of absolute space, for reasons not anticipated by Newton's contemporary critics. Its starting point was a critical questions about the law of inertia: relative to what is the motion of a free particle uniform and rectilinear? If the answer is "absolute space," then the law would appear to be something other than an empirical claim, for no one can observe the trajectory of a particle relative to absolute space. Two quite different answers to the question were offered in 1870, in the form of revised statements of the law of inertia. Carl Neumann proposed that when we state the law, we must suppose that there is a body somewhere in the universe – the "body Alpha" – with respect to which the motion of a free particle is rectilinear, and that there is a time-scale somewhere relative to which it is uniform . Ernst Mach claimed that the law of inertia, and Newton's laws generally, implicitly appeal to the fixed stars as a spatial reference-frame, and to the rotation of the earth as a time-scale; at least, he held, such is the basis for any genuine empirical content that the laws have. The notion of absolute space, it followed, was only an unwarranted abstraction from the practice of measuring motions relative to the fixed stars.

Mach's proposal had the advantage of a clear empirical motivation; Neumann's "body Alpha" seemed no less mysterious than absolute space, and almost sounds comical to the modern reader. But Neumann's discussion of a time-scale was somewhat more fruitful. He noted that the law of inertia defines a time-scale: equal intervals of time are those in which a free particle travels equal distances. He also noted, however, that this definition is quite arbitrary. For, in the absence of a prior definition of equal times, any motion whatever can be stipulated to be uniform. It is no help to appeal to the requirement of freedom from external forces, since the free particles presumably are known to us only by their uniform motion. We have a genuine empirical claim only when we state of at least two free particles that their motions are mutually proportional; equal intervals of time can then be defined as those in which two free particles travel mutually proportional distances.

THE EMERGENCE OF THE CONCEPT OF INERTIAL FRAME

Neumann's definition of a time-scale directly inspired Ludwig Lange's conception of "inertial system," introduced in 1885 . An inertial coordinate system ought to be one in which free particles move in straight lines. But any trajectory may be stipulated to be rectilinear, and a coordinate system can always be constructed in which it is rectilinear. And so, as in the case of the time-scale, we cannot adequately define an inertial system by the motion of one particle. Indeed, for any two particles moving anyhow, a coordinate system may be found in which both their trajectories are rectilinear. So far the claim that either particle, or some third particle, is moving in a straight line may be said to be a matter of convention. We must define an inertial system as one in which at least three non-collinear free particles move in noncoplanar straight lines; then we can state the law of inertia as the claim that, relative to an inertial system so defined, the motion of any fourth particle, or arbitrarily many particles, will be rectilinear. The notions of inertial system and Neumann's time-scale, which Lange called an "inertial time-scale," may be combined as follows: relative to a coordinate system in which three free particles move in straight lines and travel mutually-proportional distances, the motion of any fourth free particle will be rectilinear and uniform. The questionable Newtonian concepts of absolute rotation and acceleration, Lange proposed, could now be replaced by the concepts of "inertial rotation" and "inertial acceleration," i.e. rotation and acceleration relative to an inertial system and inertial time-scale.

By Newton's first law, a particle not subject to forces travels equal distances in equal times. But which particles are free of forces? This might appear to be a matter of convention.

See both figures below.

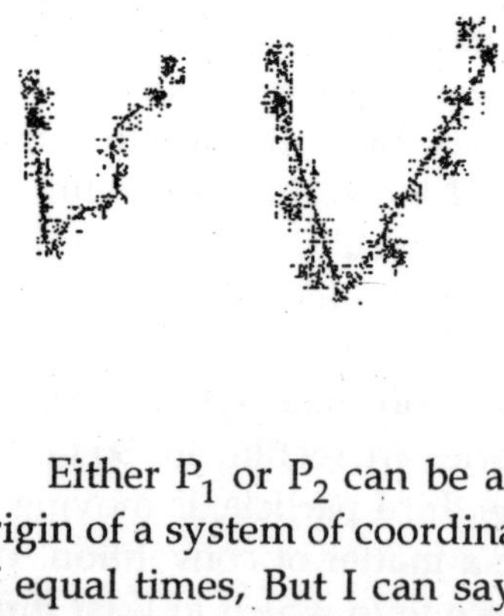

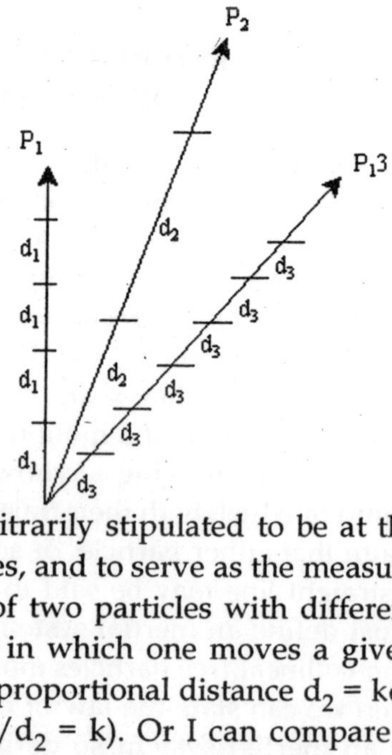

Either P_1 or P_2 can be arbitrarily stipulated to be at the origin of a system of coordinates, and to serve as the measure of equal times, But I can say of two particles with different velocities: in intervals of time in which one moves a given distance d_1, the other moves a proportional distance $d_2 = kd_1$ (where k is a constant; i.e., $d_1/d_2 = k$). Or I can compare a particle to a freely rotating planet: in intervals of time through which the planet rotates through equal angles, the particle moves equal distances.

An inertial system is a coordinate system with respect to which three free particles, projected from a single point and moving in non-coplanar directions, move in straight lines and travel mutually-proportional distances. The law of inertia then states that relative to any inertial system, any fourth free particle will move uniformly.

At about the same time, apparently unaware of the work of Mach, Neumann, and Lange, James Thomson expressed the content of the law of inertia, and the appropriate frame of reference and time-scale ("dial-traveller"), somewhat differently:

For any set of bodies acted on each by any force, a reference frame and a reference dial-traveller are kinematically possible, such that relatively to them conjointly, the motion of the mass-centre of each body, undergoes change simultaneously with any infinitely short element of the dial-traveller progress, or with any element during which the force on the body does not alter in direction nor in magnitude, which change is proportional to the intensity of the force acting on that body, and to the simultaneous progress of the dial-traveller, and is made in the direction of the force.

More simply, an inertial reference-frame is one in which Newton's second law is satisfied, so that every acceleration corresponds to an impressed force. Thomson did not reject the term "absolute rotation," holding instead that it has to be understood as rotation relative to a reference frame that satisfies his definition. The definition does not express, as Lange's does, the degree of arbitrariness involved in the construction of an inertial system by means of free particles. Moreover, like Lange's, it leaves out a crucial condition for an inertial system as we understand it: all forces must belong to action-reaction pairs. Otherwise we could have, as on a rotating sphere, merely apparent (centrifugal) forces that are, by definition, proportional to mass and acceleration, and so the rotating sphere would satisfy Thomson's definition. Therefore the definition needs to be completed by the stipulation that to every action there is an equal and opposite reaction.

But, so completed, Thomson's definition has two advantages over Lange's. First, by appealing to Newton's second law instead of his first, it shows that we can apply the notion of inertial frame without having to consider the question whether there really are any free particles in nature. Second, it exhibits more clearly an essential point about the relation between the laws of motion and the inertial frames: that the laws assert the existence of at least one inertial frame. The original question, "relative to what frame of reference do the laws of motion hold?" is revealed to be wrongly posed. For the laws of motion essentially determine a class of reference frames, and (in principle) a procedure for constructing them. For the same reason, a skeptical question that is still commonly asked about the laws of motion — why is it that the laws are true only relative to a certain choice of reference frame? — is also wrongly posed. If Newton's laws are true, then we can construct an inertial frame; their truth doesn't depend on our ability to construct such a frame in advance.

INERTIAL FRAMES IN THE 20TH CENTURY: SPECIAL AND GENERAL RELATIVITY

INERTIAL FRAMES IN NEWTONIAN SPACETIME

By the early years of the 20th century, this notion of inertial system seems to have been widely accepted, even if the specific works of Lange and Thomson were largely forgotten; in writing "On the electrodynamics of moving bodies" in 1905, Einstein took it to be obvious to his readers that classical mechanics does not require a single privileged frame of reference, but an equivalence-class of frames, all in uniform motion relative to each other, and any of which "the equations of mechanics hold good." Two inertial frames with coordinates (x, y, z, t) and (x^2, y^2, z^2, t^2) are related by the Galilean transformations,

$$x^2 = x - vt$$

$$y^2 = y$$

$$z^2 = z$$

$$t^2 = t$$

(assuming that the x axis is defined to be the direction of their relative motion). These transformations clearly preserve the invariant quantities of Newtonian mechanics, i.e. acceleration, force, and mass (and therefore time, length, and simultaneity). As far as Newtonian mechanics was concerned, then, the problem of absolute motion was completely solved; all that remained was to express the equivalence of inertial frames in a simpler geometrical structure.

The lack of a privileged spatial frame, combined with the obvious existence of privileged states of motion – paths defined as rectilinear in space and uniform with respect to time – suggests that the geometrical situation ought to be regarded from a four-dimensional spatiotemporal point of view. The structure defined by the class of inertial frames can be captured in the statement that spacetime is a four-dimensional affine space, whose straight lines (geodesics) are the trajectories of particles in uniform rectilinear motion.

Inertial Trajectories as Straight Lines of Spacetime

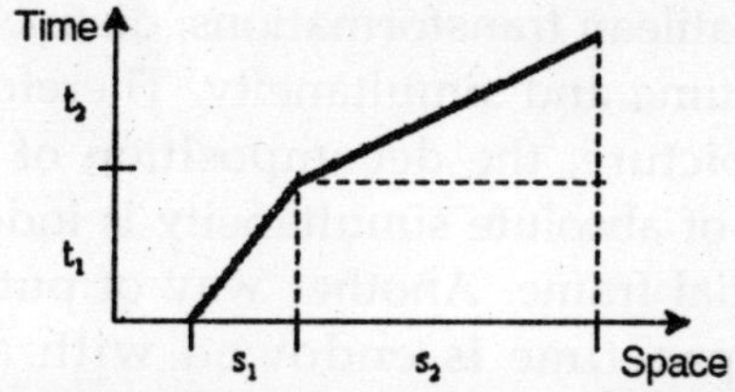

That is, spacetime is a structure whose automorphisms – the Galilean transformations that relate one inertial frame to another – are equivalent to affine transformations: they take straight lines into straight lines (i.e. an inertial motion in one inertial frame will be an inertial motion in any other inertial frame, and likewise for an accelerating or rotational motion), and parallel lines into parallel lines (i.e. uniformly-moving particles or observers who are relatively at rest in one frame will also be relatively at rest in another). (Refer Stein, Ehlers and Friedman for further explanation.) An inertial frame can be characterized as a family of parallel straight lines "filling" spacetime, representing the possible trajectories of a family of free particles that are relatively at rest.

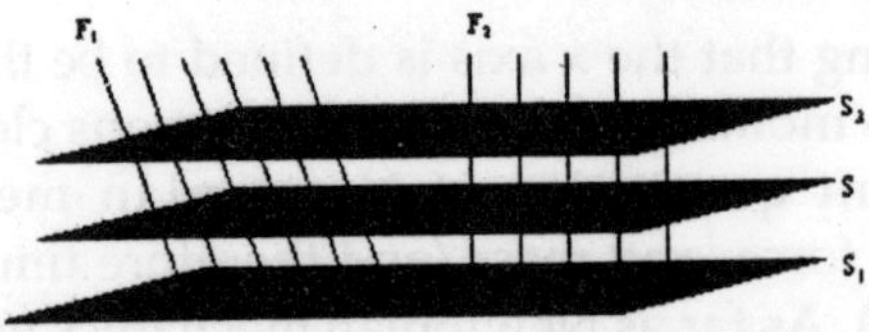

Each of these families of straight lines, F_1 and F_2, represents the trajectories of a family of free particles that are relatively at rest, and therefore each defines an inertial frame. Relative to each other, the frames defined by F_1 and F_2 are in uniform motion.

Each of the surfaces S is a "hypersurface of absolute simultaneity" representing all of space at a given moment; evidently (given the Galilean transformations) two inertial frames will agree on which events in spacetime are simultaneous.

From this we can see that the assertion that an inertial frame exists imposes a global structure on spacetime; it is equivalent to the assertion that spacetime is flat. As we can see from the Galilean transformations, distinct inertial frames will agree on time and simultaneity. Therefore, in the four-dimensional picture, the decomposition of spacetime into hypersurfaces of absolute simultaneity is independent of the choice of inertial frame. Another way of putting this is that Newtonian spacetime is endowed with a projection of spacetime onto time, i.e. a function that identifies spacetime points that have the same time-coordinate. Similarly, absolute space arises from a projection of spacetime onto space, i.e. a function that identifies spacetime points that have the same spatial coordinates.

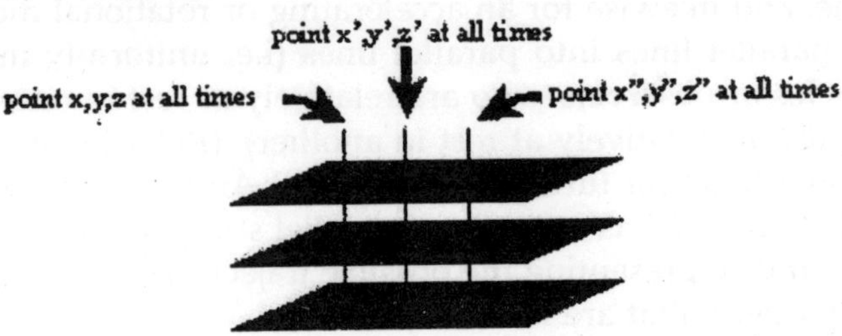

The relation of simultaneity "decomposes" spacetime into 3-dimensional pieces, each representing "all of space at a given time," by projecting spacetime onto time, i.e., by identifying spacetime points that have the same time coordinates.	Similarly, one can think of the notion of "same place" as projecting spacetime onto space, i.e., by identifying spacetime points that have the same spatial coordinates; each of the trajectories thus singled out represents "a given place at all times."

But this latter projection is arbitrary: while it assumes that we can identify the same time at different spatial locations, Newtonian mechanics provides no physical way of identifying the same spatial point at different times. Thus the equivalence of inertial frames can be thought of as the arbitrariness of the projection of spacetime onto space, any such projection being, essentially, the arbitrary choice of some particular inertial frame as a rest-frame.

Here is a spacetime diagram of motions relative to the inertial frame in which O_1, O_2, and P are at rest. This can be seen as arising from the projection of each of their inertial trajectories onto a single point of space., Here is the same situation viewed from an inertial frame in which O_3 and P_2 are at rest. Now O_1, O_2, and P are in uniform motion.

O_1 and O_2 are at rest, O_3 is at rest

O_3 is in uniform motion, O_1 and O_2 are in uniform motion

O_4 is accelerating any old way, O_4 is accelerating any old way

O_5 and O_6 are revolving around their common centre of gravity P, which is at rest, O_5 and O_6 are revolving around their common centre of gravity P, which is in uniform motion

O_7 and O_8 are revolving around their centre of gravity P_2, which is in uniform motion., O_7 and O_8 are revolving around their centre of gravity P_2, which is at rest.

THE CONFLICT BETWEEN GALILEAN RELATIVITY AND MODERN ELECTRODYNAMICS

By the time that this representation of the Newtonian spacetime structure was developed, however, the Newtonian conception of inertial frame had been essentially overthrown. First, 19th-century electrodynamics raised again the question of a privileged frame of reference: the conception of light as an electromagnetic wave in the ether implied that the rest-frame of the ether itself should play a distinguished role in electrodynamical phenomena. On the one hand, physicists such as Maxwell and Lorentz were careful to point out that velocity relative to the ether was not equivalent to absolute velocity, and that the state of motion of the ether itself was necessarily unknown — in other words, that this conception of light did not violate the classical principle of relativity. On the other hand, the existence of such a preferred frame made the equivalence of inertial frames correspondingly less interesting, even if it was true in principle. This is why the appearance of the idea of inertial frame in the 1880's, as I suggested earlier, was not of pressing physical interest to the majority of physicists, and seemed to be a mere philosophical sidelight. The attempts to measure the effects of motion relative to the ether commanded considerably more attention.

Second, the abandonment of the ether — following the failure of attempts to measure velocity relative to the ether and, more generally, the apparent independence of all electrodynamical phenomena of motion relative to the ether — did not vindicate the Newtonian inertial frame, but required a dramatically revised conception. Special relativity might be said to have applied the relativity principle of Newtonian mechanics to Maxwell's electrodynamics, by eliminating the privileged status of the rest-frame of the ether and admitting that the velocity of light is independent of the motion of the source. As Einstein expressed it, "the same laws of electrodynamics and optics will be valid for all frames of reference for which the equations of mechanics hold good." But as Einstein also pointed out, the invariance of the velocity of light and the principle of relativity, at least in its Galilean form, are incompatible. It simply makes no sense, according to Galilean relativity, that any velocity should appear to be the same in inertial frames that are in relative motion.

SPECIAL RELATIVITY AND LORENTZ INVARIANCE

Einstein solved this difficulty through his analysis of simultaneity: frames in relative motion can agree on the velocity of light only if they disagree on simultaneity; only the relativity of simultaneity makes possible the invariance of the velocity of light. This means that the transformations between inertial frames that preserve the velocity of light will not preserve simultaneity. These are the Lorentz transformations:

$$x' = \frac{x - vt}{\sqrt{1 - \frac{v'}{e'}}} \qquad y' = y \qquad z' = z \qquad t' = \frac{t - \frac{vx}{c^2}}{\sqrt{1 - \frac{v^2}{c^2}}}$$

Evidently these transformations do not preserve length and time, and so the invariant quantities of Newtonian mechanics, which presuppose invariant measures of length and time, must now depend on the choice of inertial frame. By the same token, the notions of force, mass, and acceleration can no longer be appealed to in the definition of an inertial frame. The definition must instead appeal to the invariant quantities of electrodynamics: an inertial frame is one in which light travels equal distances in equal times in arbitrary directions. What seems impossible, from the point of view of Galilean relativity, is that a frame that moves uniformly relative to such a frame should also satisfy the definition. But that, again, rests on the assumption that two inertial frames will have a common measure of simultaneity. If, as Einstein asserts, the only reasonable definition of simultaneity is one provided by light signals, then there is no determination of simultaneity that will give the same results in different inertial frames. The spacetime structure that is implied by special relativity is thus an affine space, like Newtonian spacetime, but it is not objectively divided into hypersurfaces of absolute simultaneity; the sets of simultaneous events for any inertial frame are the hyperplanes orthogonal to the trajectories that determine that frame. In other words, the choice between two inertial frames determines a choice between two distinct divisions of spacetime into space and time.

Simultaneity and Reference-Frames

The details of Einstein's argument and the structure of Minkowski spacetime can be found elsewhere . Here only one more point is worth making. It could be argued that Einstein's and Lorentz's view are completely equivalent. That is, we could assume that there is indeed a privileged frame of reference, and that the apparent invariance of the velocity of light is explained by the effects on bodies of their motion through the ether (the Lorentz contraction and time dilation). This purported distinction between empirically indistinguishable frames has often been criticized on straightforward methodological grounds, but it could be (and surely has been) argued that it is more intuitively plausible than the relativity of simultaneity. After all, knowing that (as Einstein showed) the Lorentz contraction can be derived from the invariance of the velocity of light does not, by itself, entitle us to say which of the two is the more convincing starting-point.

This is why it is so important that Einstein's 1905 paper begins with a critical analysis of the entire notion of a frame of reference. It is tacitly assumed by Lorentz's theory, and classical electrodynamics generally, that we have a reference-frame in which we can measure the velocity of light. But how is such a reference-frame determined? The distances between points in space can only be determined if it is possible to determine which events are simultaneous. In practice this is always done by light-signalling, if only in the informal sense that we identify simultaneous events when we see them at the same time. But if the spatial frame of reference is determined by light-signals, and is then to be used to measure the speed of light, we would appear to be going in a circle; the underlying assumption must be that, while light-signalling is useful and practical, it is not essential to the definition of simultaneity, and that there is a fact of the matter about which events are simultaneous that is independent of this method of signalling. This assumption was actually made explicit by James Thomson. He recognized — alone, apparently, before Einstein — that the measurement of distance involves

> the difficulty as to imperfection of our means of ascertaining or specifying, or clearly idealizing, simultaneity at distant places. For this we do commonly use signals by sound, by light, by electricity, by connecting wires or bars, and by various other means. The time required in the transmission of the signal involves an imperfection in human powers of ascertaining simultaneity of occurrences at distant places. It seems, however, probably not to involve any difficulty of idealizing or imagining the existence of simultaneity. Probably it may not be felt to involve any difficulty comparable to that of attempting to form a distinct notion of identity of place at successive times in unmarked space.

In other words, Thomson assumed that it was not a difficulty in principle, like the difficulty of determining rest in absolute space. But Einstein showed that it was precisely the same kind of difficulty, and that determinations of simultaneity involve reference to an arbitrary choice of reference-frame, just as much as determinations of velocity. Einstein's conclusion is, of course, entirely contingent on the empirical facts of electrodynamics; it could have been avoided if there were in nature a useful signal of some kind whose transmission would provide a criterion of absolute simultaneity, so that the same events would be determined to be simultaneous in all inertial frames. Or, experiments might have been able to reveal the dependence of the velocity of light on the state of motion of the source. Then synchronization by light-signals could still have been regarded as a mere practical substitute for a notion of absolute simultaneity that stood on independent grounds, empirically as well as conceptually. But as Einstein saw, because of the apparent independence of the velocity of light of the motion of the source, even "idealizing or imagining the existence of simultaneity" involves light-signaling more essentially than anyone could have realized. Unless some other criterion of simultaneity is provided, therefore, the establishment of a spatial frame of reference involves light-signaling in an essential way. In the absence of such a criterion the speed of light cannot be, as Lorentz supposed, empirically measured against the background of an inertial frame; in that case the only empirically sound definition of an inertial frame is the one that appeals to the speed of light.

FROM SPECIAL RELATIVITY AND LORENTZ INVARIANCE TO GENERAL RELATIVITY AND GENERAL COVARIANCE

It may seem surprising that, after this insightful analysis of the concept of inertial frame and its role in electrodynamics, Einstein should have turned almost immediately to call that concept into question. But he had a compelling combination of physical and philosophical motives to do so. On the physical side, he realized (along with many others) that special relativity would require some fundamental revision of the Newtonian theory of gravity. On the philosophical side, he became convinced, largely by his reading of Mach (1883), that the central role of inertial frames was an "epistemological defect" that special relativity shared with Newtonian mechanics. Only relative motions are observable, yet both of these theories purport to identify a privileged state of motion and use it to explain observable effects (such as centrifugal forces). Coordinate systems are not observable, yet both of these theories assign a fundamental physical role to certain kinds of coordinate system, namely, the inertial systems. In either theory, inertial coordinates are distinguished from all others, and the laws of physics are said to hold only relative to inertial coordinate systems. In an epistemologically sophisticated theory, both of these problems would be solved at once: the new theory would only refer to what is observable, which is relative motion; it would admit arbitrary coordinate systems, instead of confining itself to a special class of system.

Why, after all, should any genuine physical phenomenon depend on the choice of coordinate system?

Another way of putting the same point is to say that, in Newtonian mechanics and special relativity, rotation is "absolute" because the transformations between inertial frames (Galilean or Lorentzian) preserve rotational states. Thus the "absoluteness" of rotation arises precisely from singling out one type of frame, by one type of transformation, instead of allowing arbitrary transformations and arbitrary frames. Einstein held that this epistemological insight had a natural mathematical representation in the principle of general covariance, or the principle that the laws of nature are to be invariant under arbitrary coordinate transformations. More precisely, what this means is that coordinate transformations are no longer required (as in the affine spaces of Newtonian mechanics and special relativity) to take straight lines to straight lines, but only to preserve the smoothness of curves (i.e. their differentiability). The general theory of relativity was intended to be a generally covariant account of spacetime, and its general covariance was intended to express the general relativity of motion. And the theory came into being because Einstein perceived a deep connection between this project and that of finding a relativistic theory of gravitation.

THE EQUIVALENCE OF INERTIA AND GRAVITY

The philosophical motivations and implications of Einstein's view are dealt with elsewhere. We will consider here only the bearing of general relativity on the notion of an inertial frame. It is questionable whether Einstein succeeded in establishing the general relativity of motion, but it is clear that general relativity undermines the concept of inertial frame in important respects. This arises from the equivalence principle: that inertial mass – the quantity that enters into Newton's second law, and that is a measure of a body's resistance to acceleration – is equivalent to gravitational mass, the quantity that enters into Newton's law of universal gravitation. A more empirical way of expressing it is that all bodies fall with the same acceleration in the same gravitational field, or, the trajectory of a body in a given gravitational field will be independent of its mass and composition. This is the principle that Newton tested by constructing pendulums with wooden boxes as their bobs, which he would fill with different materials in order to see whether those differences made a difference to the speed of falling; they didn't. Eötvös made more precise tests in the late 19th century, and established the principle to much greater accuracy; these are the results on which Einstein would have relied. Newton also tested the principle for bodies whose masses differ greatly, by observing that Jupiter and its four moons all received precisely the same acceleration from the sun's gravitational field.

The equivalence principle suggests, however, that a freely-falling frame of reference is physically indistinguishable from an inertial frame. Newton had already noticed this, and indeed he stated it, more or less, in Corollary VI to the laws of motion:

If bodies are moving in any way whatsoever with respect to one another and are urged by equal accelerative forces along parallel lines, they will all continue to move with respect to one another in the same way as they would if they were not acted on by those forces.

For example, he was able to treat the system of Jupiter and its moons as if it were (nearly) at rest or moving uniformly in a straight line, because the attractive force of the sun acts (almost) equally on every part of the system.

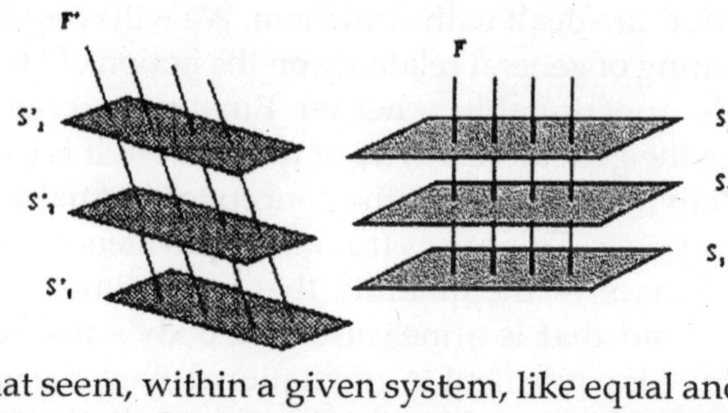

What seem, within a given system, like equal and parallel accelerations may be, on a larger scale, unequal and converging on some distant massive object; e.g., the system of Jupiter and its moons is falling toward the sun, but "locally" the accelerations are very nearly equal and parallel, and may therefore be neglected.

He even applied this reasoning to the entire solar system, in order to justify treating it as an isolated system: if there were any outside force acting on it, it must have been acting more or less equally and in parallel directions on all parts of the system.

> It may be alleged that the sun and planets are impelled by some other force equally and in the direction of parallel lines; but by such a force (by Corollary. VI of the Laws of Motion) no change would happen in the situation of the planets to one another, nor any sensible effect follow; but our business is with the causes of sensible effects. Let us, therefore, neglect every such force as imaginary and precarious, and of no use in the phenomena of the heavens.

Now, it is a familiar fact that in an orbiting spacecraft, bodies behave as if no forces were acting on any of them (as if they were "weightless"), because the attraction of the earth acts equally on all of them. But these phenomena are not, by themselves, evidence that no phenomena are capable of distinguishing an inertial frame from a falling frame. Einstein was willing to generalize the equivalence principle, and to conclude that the classical idea of a distinguished class of frames of reference has no physical basis. Any frame that we might regard as inertial might be, for all we can tell by experiment, in free fall. By the same token, any frame that is uniformly accelerating is indistinguishable from one that is at rest in a uniform gravitational field. Suppose that you are in a box at rest on the earth; you and everything in the box, by the equivalence principle, will be accelerated downward with the acceleration g (= 9.8 metres/second/second). Now suppose that the box itself is in empty, gravity-free space, but accelerating upward (i.e. in the direction of its roof) with the acceleration -g. Obviously, because of their inertia, bodies in the box, including your own, will exert the same force — have the same "weight" — on the floor as if the box were at rest and sitting on the earth.

THE EQUIVALENCE PRINCIPLE AND GENERAL COVARIANCE

To get a clearer idea of the physical significance of the equivalence principle, and its connection with general covariance, consider the Newtonian procedure for analyzing motion in the solar system, here sketched very roughly:

1. Determine the accelerations of all the planets relative to the fixed stars.
2. Using the laws of motion, their corollaries, and all the propositions proved from these in Book I of Principia, derive from the accelerations the forces needed to produce them; in particular, derive from the orbits the centres of those orbits, and the masses of the bodies needed to produce those forces. This crucially involves the law of action and reaction, for otherwise it would be impossible to break down the total acceleration of any planet into the components contributed by particular other planets; the earth's acceleration, for example, is the sum of its accelerations toward all the other planets, and each individual acceleration is part of an action-reaction pair involving some other planet.
3. When we understand the mutual interactions among the planets, we are in a position to estimate their relative masses. In Newton's case, this was necessarily restricted to the planets with satellites, because only in those cases could he compare the accelerations they determine at given distances and so deduce the differences in mass. By this reasoning he estimated the ratios of the Sun's mass to those of Jupiter (1067 to 1), Saturn (3021 to 1), and the earth, and was able to calculate that the centre of mass of the entire solar system would never be more than one solar diameter from the centre of the Sun.
4. Having found the centre of mass, we have in principle determined an inertial frame: by Corollary

IV to the laws of motion, the centre of mass will be at rest or moving uniformly in a straight line. That is, the mutual actions of the bodies in the system will not change the state of motion of the centre of mass. And having determined an inertial frame, we are in a position to say that the accelerations relative to the centre of mass frame are the true accelerations.

One might think that the problem of relativity arises right from the start: the reliance on the fixed stars already seems to introduce an arbitrary assumption that threatens to vitiate Newton's procedure as an account of the true motions. But the framework of the fixed stars, initially just taken for granted, turns out to be justified in the course of the analysis. If it turns out that all the accelerations relative to the fixed stars can be analyzed into action-reaction pairs involving bodies within the system, leaving no "leftover" accelerations that need to be traced to some yet-unknown influence, then we can conclude that the stars are a suitable (sufficiently inertial) frame of reference after all. (By the later 19th century, observations became sufficiently precise to reveal that there is in fact a leftover acceleration, namely the famous extra precession of Mercury. But that could not affect Newton's analysis in 1687.) In contrast, had we chosen the earth as a frame of reference, we would find that there are accelerations relative to this frame – e.g. Coriolis and centrifugal accelerations – that don't satisfy the law of action and reaction.

The Extension of the Relativity Principle

The relativistic aspect of this situation arises from the equivalence principle. Newton's Corollary VI said that the inertial frame we construct by this procedure is effectively indistinguishable from one in which all the bodies are undergoing equal and parallel accelerations caused by some force that acts equally on all of them; the equivalence principle asserts that gravity is such a force. In following the Newtonian procedure for constructing an inertial frame, we have constructed a frame which might be, for all we can determine empirically, falling in the gravitational field of some other system. This means that the accelerations relative to this frame cannot be known to be the "true accelerations"; they may be accelerations relative to a freely-falling trajectory just in case the centre of mass is itself freely falling, in which case they have to be added to the gravitational acceleration of the centre of mass before we can arrive at the true accelerations. But the acceleration of the centre of mass may have to be added to some larger acceleration – and so on. This means that we can't know the true strength of the gravitational field by observing the motions in this frame. The only hope of doing so would be to include all the mass in the universe in one dynamical system; if we knew the centre of mass of the entire universe, we could rule out the possibility that something else is exerting an accelerative force, since by hypothesis there would be nothing else.

We can see the significance of this more clearly by looking at the equations of motion (in a very simplified form). Newton's equation of motion for a particle subject to no force asserts that it moves uniformly, with zero acceleration. Obviously, in a gravitational field, the particle's acceleration will depend on the field. In effect, we are accounting for the trajectory of the falling particle by "decomposing" it into two parts, the part determined by its natural tendency to move uniformly in a straight line, and the part contributed by the gravitational field. But by the analysis of the equivalence principle, determining the inertial part – and therefore determining the gravitational part – depends on our assumption that the centre of mass frame is inertial rather than freely falling. And this assumption is arbitrary; that is, it amounts to an arbitrary choice of the coordinate system in which we define the equation of inertial motion. This implies that the gravitational field depends on the coordinate system in precisely the same way.

The principle of general covariance, then, acquires its physical significance in conjunction with the equivalence principle. By itself, it says that the geometrical structures of spacetime don't depend on the coordinates in which we express them, or on the set of points that we may think comprises spacetime. This is an important principle, but it doesn't recommend general relativity over other theories, since special relativity and Newtonian mechanics also involve spacetime structures that can be defined in a generally-covariant way, through the same kinds of coordinate-independent mathematical objects that we use in general relativity. Combined with the equivalence principle, however, it implies that a central Newtonian idea – that gravity is a force causing deviations from uniform rectilinear motion – is based on an arbitrary choice of coordinates. For a trajectory that satisfies all empirical criteria for being inertial in a particular frame of reference – e.g. the trajectory of the centre of mass in our example – may be freely falling relative to some other trajectory that satisfies the same criteria. By contrast, a freely-falling trajectory is a freely falling trajectory in any coordinate system; it is only the decomposition of it into its inertial and gravitational parts that will be different in different coordinate systems.

From Inertial Frames to Curved Spacetime

General covariance is thus not an argument against privileged states of motion, as Einstein had hoped it would be. It is an argument that the privileged states of motion should not be mere artifacts of our choice of coordinates, i.e. that they should be coordinate-independent. Precisely what this means depends, then, on what physical means we have at our disposal to identify states of motion other than by simply setting down coordinates. Combined with the equivalence principle, it is an argument for regarding gravitational free-fall as the privileged state of motion, rather than as a forced deviation from the privileged state of motion. And in this way it provides an argument for spacetime curvature. As we saw, in Newtonian and Minkowski spacetime the inertial trajectories are, by definition, the straight lines or geodesics of spacetime. And the flatness of spacetime consists in the fact that these geodesics behave like straight lines in a flat space or surface: parallel geodesics remain parallel, and non-parallel geodesics do not accelerate relative to one another. (In any inertial frame, the motion of any other inertial frame appears uniform.) By the equivalence principle, however, free-fall trajectories satisfy all empirical criteria for being inertial trajectories, and so the distinction between the two types of trajectory depends on the mere choice of coordinates. General covariance suggests, then, that the free-fall trajectories ought to be identified as the inertial trajectories – and therefore, as the geodesics of spacetime. But if free-fall trajectories are the geodesics of spacetime, then spacetime is curved. For the free-fall trajectories exhibit relative accelerations, and the relative acceleration of geodesics is a defining characteristic of curved geometry. The curvature of the earth's surface, for example, is revealed in the fact that geodesics that begin in parallel directions can begin to approach one another – for example, two lines of longitude can both be perpendicular to the equator, but converge on one another as they approach the poles. And since the relative accelerations of falling bodies depend on the distribution of mass, as we already knew from Newton's theory, we now conclude not only that spacetime is curved, but that its curvature is determined by the distribution of mass.

The curvature of spacetime, finally, determines the status of inertial frames in general relativity. The statement that all reference-frames, rather than just inertial frames, are equivalent is a misleading way of describing the situation; rather, the variable curvature of spacetime makes the imposition of a global inertial frame impossible. So the status of the latter is like the status of a plane rectangular coordinate system on the surface of the earth. Over a sufficiently small area, the coordinate plane may be a good approximation to the surface, but over increasingly large areas it diverges increasingly from the contours of the earth. And if two such coordinate systems, with their origins at different points on the earth, are extended until they meet, they will be seen to be "disoriented" relative to one another. In contrast, a flat plane can be so coordinatized, and coordinate systems originating at different points can be smoothly combined into one system.

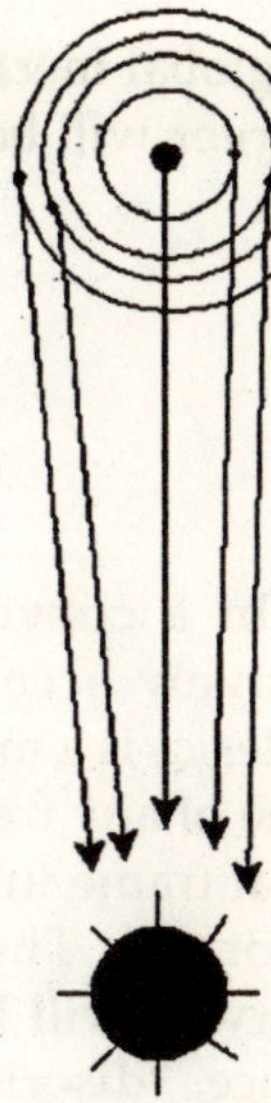

Similarly, in the affine spaces of Newtonian and special-relativistic physics, any inertial coordinate system can be extended over the whole of spacetime. And in any system so extended, the trajectory of every other inertial observer will be a uniform rectilinear motion. But if spacetime is variably curved, according to the distribution of mass and energy, local inertial systems will be "disoriented" relative to one another; indeed, the degree of this "disorientation" is one of the measures of curvature. And an inertially-moving – i.e. freely falling – particle will in that case be accelerating in the local inertial system of another freely-falling particle. Thus there are inertial trajectories, but no extended inertial systems.

This Cartesian coordinate system can evidently be simply "set down" over the plane below. Any coordinate system defined at any point of the plane can be smoothly extended over the entire plane.

Figure below is: "Magnified" View of Flat "Local" Coordinate Systems on a Curved Surface.

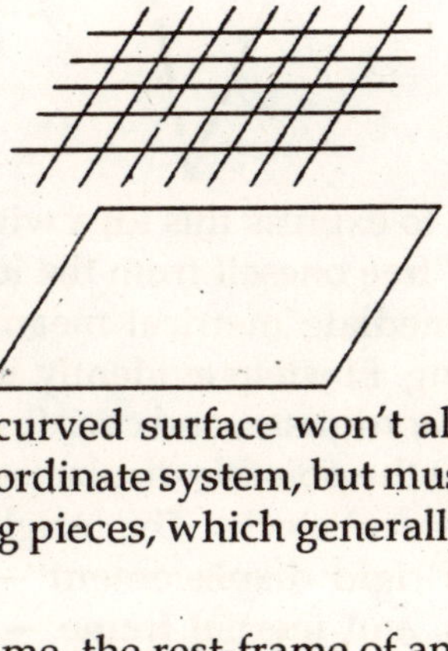

This arbitrary curved surface won't allow for the global laying down of a coordinate system, but must be coordinatized in small overlapping pieces, which generally won't be parallel to one another.

In a flat spacetime, the rest-frame of any inertial observer an be "extended" over all of spacetime in such a way that, in this global inertial frame, the trajectory of every other inertial observer will be an inertial trajectory.

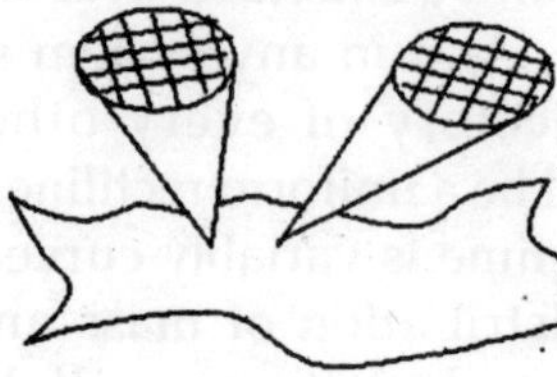

In a curved spacetime, inertial trajectories will be relatively accelerated; indeed the relative acceleration of geodesics is a measure of curvature. Therefore the local inertial frame of any freely-falling observer cannot be extended into a global frame in which all other inertial observers are moving uniformly. The inertial frames of different freely-falling observers will be, like local coordinate systems on a curved surface, "disoriented" relative to one another.

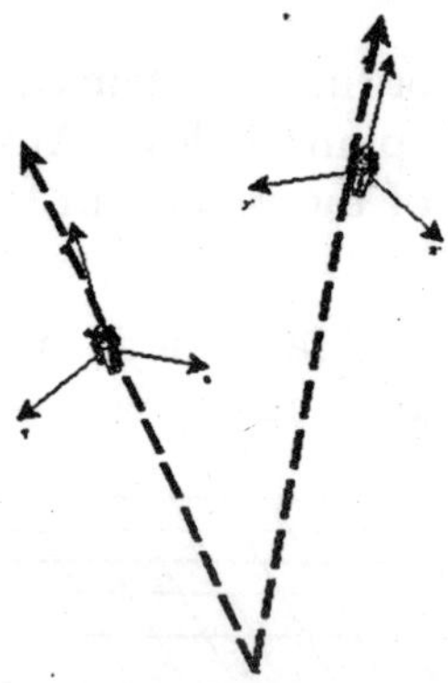

One could try to express this idea with Einstein's remark about the need to "free oneself from the idea that coordinates must have an immediate metrical meaning." But even this might be misleading. Einstein evidently was thinking that, in general relativity, coordinates, and coordinate transformations, no longer represent the possible displacements of rigid bodies or the transport of ideal clocks. The insight underlying this is that the notion of rigid displacement – therefore of rigid coordinate system, and inertial frame – imposes a priori a degree of uniformity, or symmetry, on spacetime; the displacement of bodies without change of dimension, and the transport of an ideal clock without distortion of time-intervals, requires a homogeneous space. And so rigid displacement cannot be a basic principle in a theory in which spacetime curvature varies according to the distribution of mass and energy. The possibility of a rigid displacement, and therefore the existence of an inertial frame, can only arise a posteriori, as the result of a peculiar distribution of mass-energy (for example, in a universe empty of mass and energy, or with a highly symmetrical distribution). The serious defect in the notion of inertial frame is not that it makes an arbitrary distinction among coordinate systems – for the distinction is quite as genuine as the distinction between flat and curved spacetime – but that it extends indefinitely over spacetime a structure that, in our universe, only corresponds approximately to very small regions.

Philosophical Controversy Over Absolute and Relative Motion

Leibniz, later, articulated a more general "equipollence of hypotheses": in any system of interacting bodies, any hypothesis that any particular body is at rest is equivalent to any other. Therefore neither Copernicus' nor Ptolemy's view can be true – though one may be judged simpler than the other – because both are merely possible hypothetical interpretations of the same relative motions. This principle clearly defines (what we would call) a set of reference frames, differing in their arbitrary choices of a resting point or origin, but agreeing on the relative positions of bodies at any moment and their changing relative distances through time.

For Leibniz and many others, this general equivalence was a matter of philosophical principle, founded in the metaphysical conviction that space itself is nothing more than an abstraction from the geometrical relations among bodies. In some form or other it was a widely shared tenet of the 17th-century "mechanical philosophy". Yet it was flatly incompatible with physics as Leibniz himself, and the other "mechanists," actually conceived it. For the basic programme of mechanical explanation depended essentially on the concept of a privileged state of motion, as expressed in the assumption that bodies maintain a state of rectilinear motion until acted upon by an external cause. Thus their fundamental conception of force, as the power of a body to change the state of another, likewise depended on this notion of a privileged state. This dependence was clearly exhibited in the vortex theory of planetary motion, in which every orbit was explained by the balance between the planet's inherent centrifugal tendency (its tendency to follow the tangent to the orbit) and the pressure of the surrounding medium.

For this reason, the notion of a dispute between "relativists" or "relationists" and "absolutists" or "substantivalists", in the 17th century, is a drastic oversimplification. Newton, in his controversial Scholium on space, time, and motion, was not merely asserting that motion is absolute in the face of the mechanists' relativist view; he was arguing that a conception of absolute motion was already implicit in the views of his opponents — that it was implicit in their conception, which he largely shared, of physical cause and effect. The general equivalence of reference-frames was implicitly denied by a physics that understood forces as powers to change the states of motion of bodies.

Newton therefore held that physics required the conception of absolute space, a distinguished frame of reference relative to which bodies could be said to be truly moving or truly at rest. Assuming, as both Newton and Leibniz did, that states of motion could be distinguished by their causes and effects, the distinguished status of this frame of reference is physically well founded — and metaphysically well-founded for a metaphysics that, like Newton's or Leibniz's, takes force to be a well-founded notion. On Leibniz's conception of force, in particular, a given force is required to generate or to maintain a given velocity — for objects "passively" resist motion, but maintain their states of motion only by "active" force — so that, on dynamical grounds, "every body truly does have a certain amount of motion, or, if you will, force." This implies that there is in principle a distinguished frame of reference in which the velocities of bodies correspond to their true velocities, i.e. to the amounts of moving force that they truly possess, and it implies that in any frame that is in motion relative to this one, bodies will not have their true velocities. In short, such a conception of force, if it could be applied physically, would give a precise physical application of Newton's conception of absolute space.

NON-INERTIAL FRAME OF REFERENCE

A non-inertial frame of reference does not have a constant velocity. It is accelerating. There are several ways to imagine this motion:

- The frame could be traveling in a straight line, but be speeding up or slowing down.
- The frame could be traveling along a curved path at a steady speed.
- The frame could be traveling along a curved path and also speeding up or slowing down.

Such an accelerating frame of reference is called a non-inertial frame because the law of inertia does not hold in it. That is, an object whose position is judged from this frame will seem to spontaneously change its velocity with no apparent non-zero net force acting upon it. This completely violates the law of inertia and Newton's laws of motion, since these laws claim that the only way an object can change its velocity is if an actual non-zero net force is applied to the object. Objects just do not start to move about here and there all on their own.

This is really quite easy to understand. If you are in an automobile when the brakes are abruptly applied, then you will feel pushed toward the front of the car. You may actually have to extend you arms to prevent yourself from going forward toward the dashboard. However, there is really no force pushing you forward. The car, since it is slowing down, is an accelerating, or non-inertial, frame of reference, and the law of inertia no longer holds if we use this non-inertial frame to judge your motion.

If all of this is viewed relative to the ground, it becomes clear that no force is pushing you forward when the brakes are applied. The ground is stationary and, therefore, is an inertial frame. Relative to the ground, when the brakes are applied, you continue with your forward motion, just like you should according to Newton's first law of motion. The situation is this: the car is stopping, you are not; so, you head out toward the dashboard. From your point of view rin the car it seems like you have spontaneously been pushed forward. Actually, there is no force acting on you. The imagined force toward the front of the car is a fictitious force.

A similar fictitious force can be noticed by a person in a car when it speeds up. Let us say that you are in a car at a stop light. The car is standing still. The light turns green, and the car accelerates forward. While undergoing this acceleration, the car is a non-inertial frame of reference: If the acceleration is large enough, you will feel yourself "pushed" into the seat. Actually, no force is pushing on you. Again, as viewed from the inertial frame of the ground, you are just maintaining your velocity, as you should according to Newton's first law of motion. You were still when the light was red, and you are attempting to remain still when the light turns green. However, the car started to move when the light turned green. The car actually comes up from behind you, and, using the seat, the car pushes you forward. As the seat comes forward and pushes on you, the back seat cushion compresses a bit. You may interpret this feeling as your body being pushed backward into the seat. Really, you are attempting to maintain your velocity of zero, and the seat is coming up from behind to push on you. There is no backward force. The imagined force is a fictitious force. Fictitious forces arise in non-inertial, or accelerating, frames of reference.

There are several ways to describe a non-inertial frame. Here are a few descriptions:

- A non-inertial frame of reference is a frame of reference with a changing velocity. The velocity of a frame will change if the frame speeds up, or slows down, or travels in a curved path.
- A non-inertial frame of reference is an accelerating frame of reference.
- A non-inertial frame of reference is a frame of reference in which the law of inertia does not hold.
- A non-inertial frame of reference is a frame of reference in which Newton's laws of motion do not hold.
- In a non-inertial frame of reference fictitious forces arise.

What follows here are two demonstrations that show non-inertial frames of reference. The first one is an animation of a non-inertial frame which acts like an elevator. The other shows an animation of a rotating frame of reference. Rotating frames of reference are non-inertial frames since they are following curved paths. Remember that a change in direction, which would occur along a curved path, constitutes a change in velocity, and, therefore, constitutes an acceleration. If the frame accelerates, it is a non-inertial frame.

A non-inertial frame of reference is a coordinate system which is accelerating. That is, its vector velocity is not constant. So, it is either changing its speed by speeding up or slowing down, or it is changing its direction by traveling in a curved path, or it is both changing its speed and changing its direction.

Chapter 6

The Schwarzschild Metric from Kepler's 3rd Law

The first and still the most important rigorous solution of the Einstein field equations was found by Schwarzschild in 1916. Although it's quite difficult to find exact analytical solutions of the complete field equations for general situations, the task is immensely simplified if we restrict our attention to highly symmetrical physical configurations. For example, it's obvious that the flat Minkowski metric trivially satisfies the field equations. The simplest non-trivial configuration in which gravity plays a role is a static mass point, for which we can assume the metric has perfect spherical symmetry and is independent of time. Let r denote the radial spatial coordinate, so that every point on a surface of constant r has the same intrinsic geometry and the same relation to the mass point, which we fix at r = 0. Also, let t denote our temporal coordinate. Any surface of constant r and t must possess the two-dimensional intrinsic geometry of a 2-sphere, and we can scale the radial parameter r such that the area of this surface is $4\pi r^2$. (Notice that since the space may not be Euclidean, we don't claim that r is "the radial distance" from the mass point. Rather, at this stage r is simply an arbitrary radial coordinate scaled to give the familiar Euclidean surface area.) With this scaling, we can parameterize the two-dimensional surface at any given r (and t) by means of the ordinary "longitude and latitude" spherical metric

$$(dS)^2 = r^2 (d\theta)^2 + r^2 \sin(\theta)^2 (d\phi)^2$$

Where dS is the incremental distance on the surface of an ordinary sphere of radius r corresponding to the incremental coordinate displacements $d\theta$ and $d\phi$. The coordinate θ represents "latitude", with $\theta = 0$ at the north pole and $\theta = \pi/2$ at the equator. The coordinate f represents the longitude relative to some arbitrary meridian.

On this basis, we can say that the complete spacetime metric near a spherically symmetrical mass m must be of the form

$$(d\tau)^2 = g_{tt}(dt)^2 + g_{rr}(dr)^2 + g_{\theta\theta}(d\theta)^2 + g_{\phi\phi}(d\phi)^2$$

where $g_{\theta\theta} = -r^2$, $g_{\phi\phi} = -r^2\sin(\theta)^2$, and g_{tt} and g_{rr} are (as yet) unknown functions of r and the central mass m. Of course, if we set m = 0 the functions g_{tt} and $-g_{rr}$ must both equal 1 in order to give the flat Minkowski metric (in polar form), and we also expect that as r increases to infinity these functions both approach 1, regardless of m, since we expect the metric to approach flatness sufficiently far from the gravitating mass.

This metric is diagonal, so the non-zero components of the contravariant metric tensor are $g^{\alpha\alpha} = 1/g_{\alpha\alpha}$. In addition, the diagonality of the metric allows us to simplify the definition of the Christoffel symbols to

$$\Gamma^{\alpha}_{\mu\nu} = \frac{1}{2g_{\alpha\alpha}}\left(\frac{\partial g_{\alpha\mu}}{\partial x^{\nu}} + \frac{\partial g_{\alpha\nu}}{\partial x^{\mu}} - \frac{\partial g_{\mu\nu}}{\partial x^{\alpha}}\right) \text{ (no implied summations)}$$

Now, the only non-zero partial derivatives of the metric coefficients are

$$\frac{\partial g_{\theta\theta}}{\partial r} = -2r \qquad \frac{\partial g_{\phi\phi}}{\partial r} = -2r\sin(\theta)^2 \qquad \frac{\partial g_{\phi\phi}}{\partial \theta} = -2r^2\sin(\theta)\sin(\theta)$$

along with $\partial g_{tt}/dr$ and $\partial g_{rr}/dr$, which are yet to be determined. Inserting these values into the preceding equation, we find that the only non-zero Christoffel symbols are

$$\Gamma' tr = \Gamma' rt = \frac{1}{2g_{tt}}\frac{\partial g_{tt}}{\partial r}$$

$$\Gamma^t{}_{tr} = \frac{1}{2g_{rr}}\frac{\partial g_{tt}}{\partial r} \Gamma^r{}_{rr} \qquad \frac{1}{2g_{rr}}\frac{\partial g_{rr}}{\partial r} \Gamma^r{}_{\theta\theta} \qquad \frac{r}{\partial_{rr}} \Gamma^r{}_{\phi\phi}$$

$$\frac{r din(\theta)^2}{g_{rr}}$$

$$\Gamma^\theta{}_{r\theta} = \Gamma^\theta{}_{\theta r} = \frac{1}{r} \qquad \Gamma^\theta{}_{\phi\phi} = \sin(\theta)\cos(\theta)$$

$$\Gamma^\phi{}_{\phi r} = \Gamma^\phi{}_{r\phi} = \frac{1}{r} \qquad \Gamma^\phi{}_{\phi\theta} = \Gamma^\phi{}_{\theta\phi} = \frac{1}{\tan(\theta)}$$

These are the coefficients of the four geodesic equations near a spherically symmetrical mass. We assume that, in the absence of non-gravitational forces, all natural motions (including light rays and massive particles) follow geodesic paths, so these equations provide a complete description of inertial/gravitational motions of test particles in a spherically symmetrical field. All that remains is to determine the metric coefficients g_{tt} and g_{rr}.

We expect that one possible solution should be circular Keplerian orbits, i.e., if we regard r as corresponding (at least approximately) to the Newtonian radial distance from the centre of the mass, then there should be a circular geodesic path at constant r that revolves around the central mass m with an angular velocity of w, and these quantities must be related (at least approximately) in accord with Kepler's third law

$$m = r^3 \omega^2$$

(It's interesting that the original deductions of an inverse-square law of gravitation by Hooke, Wren, Newton, and others were based on this same empirical law. If we consider purely circular motion on the equatorial plane ($\theta = \pi/2$) at constant r, the metric reduces to

$$(d\tau)^2 = g_{tt}(dt)^2 - r^2 (d\phi)^2$$

and since $dr/d\tau = 0$ the geodesic equations are simply

$$\frac{d^2t}{d\tau^2} = \frac{d^2\theta}{d\tau^2} = \frac{d^2\theta}{d\tau^2} = 0 \qquad \frac{\partial g_{tt}}{\partial r}\left(\frac{dt}{d\tau}\right)^2 = 2r\left(\frac{d\phi}{d\tau}\right)^2$$

Multiplying through by $(d\tau/dt)^2$ and identifying the angular speed ω with the derivative of ϕ with respect to the coordinate time t, the right hand equation becomes

$$\frac{\partial g_{tt}}{\partial r} = 2r\left(\frac{d\phi}{dt}\right)^2 = 2r\omega^2$$

For consistency with Kepler's Third Law we must have ω^2 equal (or very nearly equal) to m/r^3, so we make this substitution to give

$$\frac{\partial g_{tt}}{\partial r} = \frac{2m}{r^2}$$

Integrating this equation, we find that the metric coefficient g_{tt} must be of the form k - (2m/r) where k is a constant of integration. Since g_{tt} must equal 1 when m = 0 and/or as r approaches infinity, it's clear that k = 1, so we have

$$g_{tt} = 1 - \frac{2m}{r}.$$

Also, for a photon moving away from the gravitating mass in the purely radial direction we have $d\tau = 0$, and so our basic metric for a purely radial ray of light gives

$$g_{tt}\,(dt)^2 = -g_{rr}\,(dr)^2$$

Invoking the symmetry $v \leftrightarrow 1/v$, we select the factorization $g_{tt} = dr/dt$ and $g_{rr} = -dt/dr$, which implies $g_{rr} = -1/g_{tt}$. This gives the complete Schwarzschild metric

$$(d\tau)^2 = \left(1 - \frac{2m}{r}\right)(dt)^2 - \left(\frac{r}{r-2m}\right)(dr)^2 - r^2(d\theta)^2 - r^2\sin(\theta)^2(d\phi)^2$$

from which nearly all of the experimentally accessible consequences of general relativity follow.

In matrix form the Schwarzschild metric is written as

$$g_{\mu\nu} = \begin{bmatrix} \frac{r-2m}{r} & 0 & 0 & 0 \\ 0 & \frac{-r}{r-2m} & 0 & 0 \\ 0 & 0 & -r^2 & 0 \\ 0 & 0 & 0 & -r^{2\sin(\theta)^2} \end{bmatrix}$$

Now that we've determined g_{tt} and g_{rr}, we have the partials

$$\frac{\partial g_{tt}}{\partial r} = \frac{2m}{r^2} \qquad \frac{\partial g_{rr}}{\partial r} = \frac{2m}{(r-2m)^2}$$

so the Christoffel symbols that we previously left undetermined are

$$\Gamma^t{}_{tr} = \Gamma^t{}_{rt} = \frac{m}{r(r-2m)} \qquad \Gamma^r{}_{rr} = \frac{-m}{r(r-2m)}$$

$$\Gamma^r{}_{\theta\theta} = -(r-2m) \qquad \Gamma^r{}_{\phi\phi} = -(r-2m)\sin(\theta)^2$$

Therefore, the complete set of geodesic equations for the Schwarzschild metric are

$$\frac{d^2t}{d\lambda} = \frac{2m}{r(r-2m)}\left(\frac{dr}{d\lambda}\right)\left(\frac{dt}{d\lambda}\right)$$

$$\frac{d^2r}{d\lambda^2} = \frac{m(r-2m)}{r^3}\left(\frac{dt}{d\lambda}\right)^2 + \frac{m}{r(r-2m)}\left(\frac{dr}{d\lambda}\right)^2$$

$$+(r-2m)\left(\frac{d\theta}{d\lambda}\right)^2 + (r-2m)\sin(\theta)^2\left(\frac{d\phi}{d\lambda}\right)^2$$

$$\frac{d^2\theta}{d\lambda^2} = -\frac{2}{r}\left(\frac{dr}{d\lambda}\right)\left(\frac{d\theta}{d\lambda}\right) + \sin(\theta)\cos(\theta)\left(\frac{d\phi}{d\lambda}\right)^2$$

$$\frac{d^2\phi}{d\lambda^2} = -\frac{2}{r}\left(\frac{dr}{d\lambda}\right)\left(\frac{d\phi}{d\lambda}\right) + \frac{2}{\tan(\theta)}\left(\frac{d\theta}{d\lambda}\right)\left(\frac{d\phi}{d\lambda}\right)$$

There are all parametric equations, where λ denotes a parameter that monotonically varies along the path. When dealing with massive particles, which travel at sub-light speeds, we must choose λ proportional to τ, the integrated lapse of proper time along the path. On the other hand, the lapse of proper time along the path of a massless particle (such as a photon) is zero by definition, so this raises an interesting question: How is it possible to extremize the "length" of a path whose length is identically zero? Even though the path of a photon has singular proper time, the path is not singular in all respects, so we can still parameterize the path by simply assigning monotonic values of λ to the points on the path. (Notice that, since geodesics are directionally symmetrical, it doesn't matter whether λ is increasing or decreasing in the direction of travel.) An alternative approach to solving for light-like geodesics, based on Fermat's principle of least time.

We applied Kepler's Third Law as a heuristic guide to these equations of motion, but there is a certain ambiguity in the derivation, due to the distinction between coordinate time t and the orbiting object's proper time τ. Recall that we defined the angular speed ω of the orbit as dϕ/dt rather than dϕ/dτ. This illustrates the unavoidable ambiguity in carrying over Newtonian laws of mechanics to the relativistic framework. Newtonian physics didn't distinguish between the proper time along a particular path and coordinate time - not surprisingly - since the two are practically indistinguishable for objects moving at much less than the speed of light. Nevertheless, the slight deviation between these two time parameters has observable consequences, and provides important tests for distinguishing between the space geodesic approach and the Newtonian force-at-a-distance approach to gravitation. We've assumed that Kepler's Third law is exactly satisfied with respect to coordinate time t, but only approximately with respect to the orbiting object's proper time τ. It's interesting that the Newtonian free-fall formulas for purely radial paths are also applicable exactly in relativity, but only if time is interpreted as the proper time of the falling particle. Thus we can claim an exact correspondence between Newtonian and relativistic laws in each of these two fundamental cases by a suitable correspondence of the time coordinates, but no single correspondence works for both of them.

To show that the equations of motion derived above (taking τ as the parameter λ) are fully equivalent to those of Newtonian gravity in the weak slow limit, we need only note that the scale factor between r and t is so great that we can neglect any terms that have a factor of dr/dt unless that term is also divided by r, in which case the scale factor cancels out. Also we can assume that $dt/d\tau$ is essentially equal to 1, and it's easy to see that if the motion of a test particle is initially in the plane $\theta = \pi/2$ then it remains always in that plane, and by spherical symmetry this applies to all planes. So we can assume $\theta = \pi/2$ and with the stated approximations the equations of motion reduce to the familiar Newtonian equations

$$\ddot{r} = -\frac{m}{r^2} + r\omega^2 \qquad \dot{\omega} = -\frac{2}{r}\dot{r}\omega$$

where ω is the angular velocity.

LORENTZ TRANSFORMATIONS AND THE ADDITION OF VELOCITIES

On June 30, 1905 Einstein formulated the two postulates of special relativity:

The Principle of Relativity

The laws of physics are the same in all inertial frames of reference.

The Constancy of Speed of Light in Vacuum

The speed of light in vacuum has the same value c in all inertial frames of reference.

The speed of light in vacuum c (299792458 m/s) is so enormous that we do not notice a delay between the transmission and reception of electromagnetic waves under normal circumstances.

The speed of light in vacuum is actually the only speed that is absolute and the same for all observers as was stated in the second postulate.

LORENTZ TRANSFORMATIONS

A peculiar effect of Einstein's postulates is the transformation that connects space-time in two inertial frames. Such transformations are called Lorentz transformations.

The standard Lorentz transformation in the x direction is (for reference also the classical Galilei transformation is included):

$$
\begin{array}{ll}
t' = \gamma(t - v\,x/c^2) & \qquad t' = t \\
x' = \gamma(x - v\,t) & \qquad x' = x - v\,t \\
y' = y & \qquad y' = y \\
z' = z & \qquad z' = z
\end{array}
$$

where $\gamma = \gamma(v) = 1/\sqrt{1 - v^2/c^2}$ is the Lorentz factor. Note that the spatial coordinates (y and z) perpendicular to the direction of motion (x) are unchanged. In the classical limit $v/c \approx 0$, which means that $\lambda \rightarrow 1$, special relativity is equivalent with classical Newtonian mechanics. Furthermore, note that $t' \neq t$, i.e. time is relative in special relativity.

Directly from Lorentz transformations, one obtains the concepts of length contraction, time dilation, relativistic Doppler effect, and relativistic addition of velocities

LENGTH CONTRACTION

Length contraction, according to Albert Einstein's special theory of relativity, is the decrease in length observed in objects traveling at any non-zero velocity relative to an observer. These contractions only become noticeable, however, at substantial fractions of the speed of light; and the contraction is only in the direction parallel to the direction in which the observed body is travelling.

It is important to note that this effect is negligible at everyday speeds, and can be ignored for all regular purposes. It is only when an object approaches speeds on the order of 30,000,000 m/s, i.e. 1/10 of the speed of light, then it becomes important. As the magnitude of the velocity approaches the speed of light, the effect becomes dominant, as we can see from the formula:

Note that in this equation it is assumed that the object is parallel with its line of movement. Also note that for the observer in relative movement, the length of the object is measured by subtracting the simultaneously measured distances of both ends of the object. For more general conversions, see the Lorentz transformations.

An observer at rest viewing an object travelling at the speed of light would observe the length of the object in the direction of motion as zero. Among other reasons, this suggests that objects with mass cannot travel at the speed of light.

Physical Origin of Length Contraction?

Length contraction as a physical effect on bodies composed of atoms held together by electromagnetic forces was proposed by George Fitzgerald based on a paper by Oliver Heaviside in the 1880s and worked out in more detail by Hendrik Lorentz (1895, 1899) and Joseph Larmor (1897, 1900). The following quote from Larmor is indicative of the pre-Relativity view of the effect as a consequence of James Clerk Maxwell's electromagnetic theory:

> "... if the internal forces of a material system arise wholly from electromagnetic actions between the system of electrons which constitute the atoms, then the effect of imparting to a steady material system a uniform velocity of translation is to produce a uniform contraction of the system in the direction of motion, of amount $(1-v^2/c^2)^{1/2}$".

The extension of this specific result to a general result was (and is) considered "ad hoc" by many who prefer Einstein's deduction of it from the Principle of Relativity without reference to any physical mechanism. In other words, length contraction is merely a consequnce of the postulates of special relativity. To gain a little physical insight on why length contractions occur, consider what those postulates involve: by requiring the speed of light (a quantity dependent on the fundamental properties of space and time) to be invariant in all frames of reference (including ones in motion) one can appreciate that it would require the "distortion" of the measures of length and time. Apparently Lorentz did not agree to the criticism that his proposal was "ad hoc" as this draft of a letter to Einstein in 1915 shows..

"... the interpretation given by me and FitzGerald was not artificial. It was more so that it was the only possible one, and I added the comment that one arrives at the hypothesis if one extends to other forces what one could already say about the influence of a translation on electrostatic forces. Had I emphasized this more, the hypothesis would have created less of an impression of being invented ad hoc."

The Trouton-Rankine experiment in 1908 showed that length contraction of an object according to one frame, did not cause changes in the resistance of the object in its rest frame. This is in agreement with some current theories at the time (Special Relativity and Lorentz ether theory) but in disagreement with Trouton and Rankine's ideas on length contraction.

A Trigonometric Effect?

Draw a rotated cuboid in three-dimensional euclidean space E^3. The cross section is longer in the direction of the rotation than it was before the rotation. The world slab of a moving thin plate in Minkowski spacetime (with one spatial dimension suppressed) $E^{1,2}$, which is a boosted cuboid. The cross section is thinner in the direction of the boost than it was before the boost. In both cases, the transverse directions are unaffected and the three planes meeting at each corner of the cuboids are mutually orthogonal (in the sense of $E^{1,2}$ at right, and in the sense of E^3 at left).

The modern view is that the so-called "Lorentz contraction" is essentially a geometric, in fact a trigonometric phenomenon, which is analogous to something which happens when we consider parallel slices through a cuboid before and after a rotation in E^3. This is the euclidean analog of boosting a cuboid in $E^{1,2}$. In the latter case, however, we can interpret the boosted cuboid as the world slab of a moving plate.

Special relativity concerns relativistic kinematics. Poincaré transformations are a class of affine transformations which can be characterized as the transformations between alternative

Cartesian coordinate charts on Minkowski spacetime corresponding to alternative states of inertial motion (and different choices of an origin). Lorentz transformations are Poincaré transformations which are linear transformations (preserve the origin).

Lorentz transformations play the same role in Minkowski geometry (the Lorentz group forms the isotropy group of the self-isometries of the spacetime) which are played by rotations in euclidean geometry. Indeed, special relativity largely comes down to studying a kind of noneuclidean trigonometry in Minkowski spacetime, as suggested by the following table:

Three Plane Trigonometries

Trigonometry, Circular, Parabolic, Hyperbolic

Kleinian Geometry, euclidean plane, Galilean plane, Minkowski plane

Symbol, E^2, $E^{0,1}$, $E^{1,1}$

Quadratic form, positive definite, degenerate, non-degenerate but indefinite

Isometry group, E(2), E(0,1), E(1,1)

Isotropy group, SO(2), SO(0,1), SO(1,1)

Type of isotropy, rotations, shears, boosts

Cayley algebra, complex numbers, dual numbers, Minkowski numbers μ^2, -1, 0, 1

Spacetime interpretation, none, Newtonian spacetime, Minkowski spacetime

slope, tan Æ = m, tanp Æ = u, tanh Æ = v

"cosine", cos Æ = $(1+m^2)^{-1/2}$, cosp Æ = 1, cosh Æ = $(1-v^2)^{-1/2}$

"sine", sin Æ = m $(1+m^2)^{-1/2}$, sinp Æ = u, sinh Æ = v $(1-v^2)^{-1/2}$

"secant", sec Æ = $(1+m^2)^{1/2}$, secp Æ = 1, sech Æ = $(1-v^2)^{1/2}$

"cosecant", csc Æ = m^{-1} $(1+m^2)^{1/2}$, cscp Æ = u^{-1}, csch Æ = v^{-1} $(1-v^2)^{1/2}$

Lorentz–FitzGerald Contraction Hypothesis

The Lorentz-FitzGerald contraction hypothesis was proposed by George FitzGerald and independently proposed and extended by Hendrik Lorentz to explain the negative result of the Michelson-Morley experiment, which attempted to detect Earth's motion relative to the luminiferous aether.

After reading a paper by Heaviside in which was shown that electric and magnetic fields are deformed by motion, FitzGerald inferred that similarly, when a body moves through space it experiences a deformation due to motion, and that this may explain the "null result". Lorentz showed independently how such an effect might be expected based on electromagnetic theory and the electrical constitution of matter, that is, when a body moves through space its dimension parallel to the line of motion might become less by an amount dependent on its speed. If the speed of the body is v and the speed of light is c, then the contraction is in the ratio

For Earth moving in its orbit at about 30 km/s (18.5 mile/s), the contraction would amount to about one part in 200,000,000, which would be about 6 cm (2.5 inches) on the diameter of Earth. This small change accounts for Michelson and Morley's negative result by making the source of light and the mirror draw closer together when the system is moving lengthwise.

The formula itself suggests that it is impossible for the velocity of objects (v) to surpass the speed of light. Doing so results in a fraction larger than 1; thus a square root of a negative number.

Relationship to Special Relativity

Henri Poincaré was not at first entirely satisfied with FitzGerald's hypothesis. In Science and Hypothesis he commented on the Lorentz contraction:

> "Then more exact experiments were made, which were also negative; neither could this be the result of chance. An explanation was necessary, and was

forthcoming; they always are; hypotheses are what we lack the least"

The Lorentz-FitzGerald contraction effect was introduced by Lorentz in paragraph 8 of his paper "Electromagnetic phenomena in a system moving with any velocity less than that of light", 1904. The hypothesis was directed specifically towards electrons with the final intent of explaining the unexpected result of the Trouton-Noble experiment and the Michelson-Morley experiment. Lorentz does this in paragraph 10 of the same paper. Albert Einstein derived the Lorentz contraction directly from the Principle of relativity ("Zur Elektrodynamik bewegter Körper"/"On the Electrodynamics of Moving Bodies" 1905). According to Einstein, early explanation attempts including the Lorentz-Fitzgerald contraction hypothesis had been "ad-hoc".

Lorentz did not agree as can be seen from his draft letter of 1915 to Einstein:

> "I felt the need for a more general theory, as I tried to develop later, and as has actually been developed by you (and to a lesser extent by Poincaré). However, my approach was not so terribly unsatisfactory. [...] And the interpretation given by me and FitzGerald was not artificial. It was more so that it was the only possible one, and I added the comment that one arrives at the hypothesis if one extends to other forces what one could already say about the influence of a translation on electrostatic forces. Had I emphasized this more, the hypothesis would have created less of an impression of being invented ad hoc."

Lorentz later believed that relativity had introduced some doubt about whether the length contraction was apparent or real. In his view "... there can be no question about the reality of the change of length ... [rod II] will be shorter than [rod I], just as it would be if it were kept at a lower temperature ..." Lorentz, H. A. "The Michelson-Morley experiment and the dimensions of moving bodies", Time dilation and length contraction as given above describe how coordinates in two different frames are related for particular types of measurements. Time dilation occurs when observers compares the readings of two clocks in different frames of reference. Length contraction relates to the apparent decrease in the length of objects in motion relative to the observer.

We will now write down the general form of the transformation laws of Special Relativity that replace the Galilean transformation laws at high high relative velocities. They are:

Where you will recall that is the time dilation factor11.6. These remarkably simple formulas are called the Lorentz transformation laws. They contain virtually everything you need to know about special relativity. It is not necessary to study them in detail in order to understand the implications of Special Relativity. We write them down for two reasons. First, a cursory inspection will show that for low velocities, they are nearly identical to the Galilean transformation laws. Remember that if v is much less than c, is close to one if you ignore the Galilean equivalent. Moreover, the last term will be negligibly small when v is small, so that in this limit the Lorentz transformation equation, gives t = t′. This correspondence with the Galilean transformation at low velocities had to be true for Special Relativity to be consistent with our everyday experience.

The second reason for writing down the equation is that they will allow us to deduce the expression for the addition of velocities in special relativity. We will do this in the same way that we derived the simple addition of velocities in the case of Galilean transformations. Let us consider a ball that is thrown in O′ at a speed of u′. If the ball is thrown at t = t′ = 0 and x = x′ = 0 then in the O′ frame the coordinate of the ball as a function of time is just x′ = u′t′ where u′ is the velocity relative to O′. The Lorentz transformation laws tell us what the corresponding coordinates are in the unprimed frame. This immediately gives us the relationship between the velocities u′and u:

Notice that the time dilation factor completely cancels out of the right hand side of the above equation. If we now divide the top and bottom by t′, replace x′/t′ by u′ everywhere it occurs, we get the velocity transformation law:

It differs from the common sense Galilean transformation law only by the second term in the denominator, but this extra term makes a world of difference for high relative velocities. However, if both u and v are small compared to c this term is negligibly small, and the formula reduces precisely to the Galilean transformation law for the addition of velocities. However, for speeds that are significant compared to that of light, some very remarkable things happen.

For example, if instead of a ball, a beam of light is emitted by O′, at a speed u′ = c. What is the speed as measured in the frame O? In this case 11.11 gives

u = = c = c,

As required by Special Relativity, the speed of light as measured in O is c, irrespective of the speed with which O′ is moving.

We can also ask what happens if the frame O′ is moving at three quarters the speed of light, and the ball is thrown at three quarters the speed of light in the same direction. What is the net velocity of the ball relative to O? Putting u = 3c/4 and v = 3c/4 into 11.11 we find

u′ = = c = c,

The net speed is still less than c! In fact it is easy to verify that no matter how fast O′ is going (as long as it is less than c) and no matter how fast the ball is thrown (as long as it is less than c), the net velocity as measured relative to O will always be less than cas well.

There is yet another important implication of all this. In order to increase the velocity of an object you must exert a force over some finite distance. Thus, the agent exerting the force must to some extent be moving with the object that is being accelerated. We can approximate this process by a series of steps in which you accelerate the object instantaneously, then catch up to it and repeat the process. In particular, suppose you throw a ball at a speed of 3c/4, and then fire rockets so that you are moving along side. You then throw the ball again at 3c/4 relative to your new frame. The Galilean transformation law would imply that the ball is now moving relative to the initial frame at the 3c/2. However the relativistic law says that the ball is only moving at 24c/25 with respect to the initial frame. Since we are trying to approximate a process whereby you continuously accelerate the ball over a finite distance, you now fire your rockets again to catch up with the ball and then throw the ball one more time at 3c/4 relative to your new frame. In this case (v = 24c/25, 3c/4) the transformation law gives a velocity of relative to the initial frame of u = 171c/172. This is close, but still less than c. It should be clear at this stage, that no matter how often you repeat this process you will never get the ball to exceed the speed of light, although by repeating this process enough times you can get arbitrarily close. The implication is it is impossible to accelerate an object to speeds greater than that of light by exerting a finite force is over finite distance (i.e. by doing a finite amount of work). Thus, c is the ultimate speed limit in the Universe.

CURVED COORDINATES AND ACCELERATION

The equality of inertial and gravitational mass has been recognized since at least the time of Galileo, and more recently has been empirically verified to 1 part in 10^{12}. As a consequence of this equality, the free fall path of a small test particle in a gravitational field is independent of the particle's mass and composition. This fact is sometimes called the "weak equivalence principle". Einstein extrapolated this idea into what is called the "strong equivalence principle", which asserts that gravity is nothing other than the effects of the geometry of spacetime. In some modern popular accounts of general relativity the strong equivalence principle is expressed in terms of a supposed impossibility of distinguishing between inertia and gravity inside an elevator car (for example). Although these descriptions are evocative and somewhat useful for conveying the underlying idea of the strong equivalence principle - which is one of the most profound and significant in all of physics - they sometimes cause people to misconstrue the essential meaning of the principle, and to go off trying to think of ways to distinguish between various kinds of acceleration fields and gravitational fields. For example, they may cite the tidal effects of a non-uniform gravitational field as a means of distinguishing the effects of such a field from the effects of linear acceleration, thereby (they imagine) contradicting the popular descriptions of the strong equivalence principle.

Such objections fundamentally miss the point of the equivalence principle, which refers to the intrinsic rather than extrinsic qualities of gravity and acceleration. To make this clear, it's important to point out that the principle of "indistinguishability" is valid (up to some finite level of precision) only locally, which is to say, only over a sufficiently small interval of both space and time. Unfortunately, talking about "levels of precision" and "sufficiently small intervals" sometimes causes people to think that the equivalence principle is, in some sense, only approximately valid, and so they may be tempted to underestimate its significance. Of course, it's undeniably true that many of the popular expressions of the equivalence principle are only approximately true, but this shouldn't cause us to lose sight of the fact that the actual equivalence principle is precisely valid - at least as far as we can verify with current experimental techniques.

It's worth noting that, even the "weak equivalence principle", i.e., the equality of inertial and gravitational mass, is an intrinsic principle that applies only on a limiting sense. The accelerations used in establishing this correspondence can be defined with great precision only when the test objects are sufficiently small, so that they don't contribute significantly to the gravitational field in the region of interest, and therefore don't "act back" significantly on the gravitating body and upset our frame of reference. Thus, even neglecting air resistance and the rotation of the Earth, we cannot say that every object will take the same amount of time to "fall to Earth" from a given height, because that is only strictly true in the limit of infinitesimally small test particles. A real falling massive object contributes its own gravity, and actually pulls the Earth toward it, reducing the time it takes for the system to collapse.

Nevertheless, this doesn't change the fact that inertial and gravitational mass are (as far as we know) identical, because the equivalence is understood to be intrinsic. The "force" of gravity couples to all infinitesimal particles of mass in the same way, which is what makes it feasible to interpret gravity as the effect of spacetime geometry. The equivalence is exact (as far as we can tell), and this is a very profound and significant fact about the physical world, and it applies at each and every point in spacetime, but in the case of extended objects we must take account of the non-gravitational forces that are holding the particles of mass in fixed configurations, rather than allowing them to each follow geodesic paths. Thus, the equivalence principle relies on our ability to conceptually separate out gravity from the other forces of nature, which are never entirely absent, so this again involves an act of abstraction to isolate the equivalence principle in its pure and exact sense.

In developing the general theory of relativity, Einstein found that, beginning from a Minkowskian framework, the field equations of gravity follow almost uniquely from the combination of the equivalence principle and the requirement of optimal simplicity under general covariance. Thus the field equations of general relativity are virtually the embodiment of the equivalence principle. In this theory, the influence of both inertia and gravitation on a particle according to the general theory of relativity is expressed by the equations of motion

$$\frac{d^2X_\mu}{ds^2} + \Gamma^\mu{}_{\alpha\beta}\frac{dX_\alpha}{ds}\frac{dX_\beta}{ds} = 0$$

It's interesting that Einstein consistently emphasized that the Christoffel symbols $G^{m}{}_{ab}$ represent the gravitational field, in spite of the fact that this sort of comment would earn most students a failing grade in a university course on relativity. The "correct" view is that the Riemann curvature tensor $R_{\mu\nu\alpha\beta}$ represents "true" gravity, whereas the Christoffel symbols can be non-zero in perfectly flat spacetime, simply by virtue of curvilinear coordinates. To illustrate, consider a flat plane with either Cartesian coordinates x,y or polar coordinates r,θ as shown below

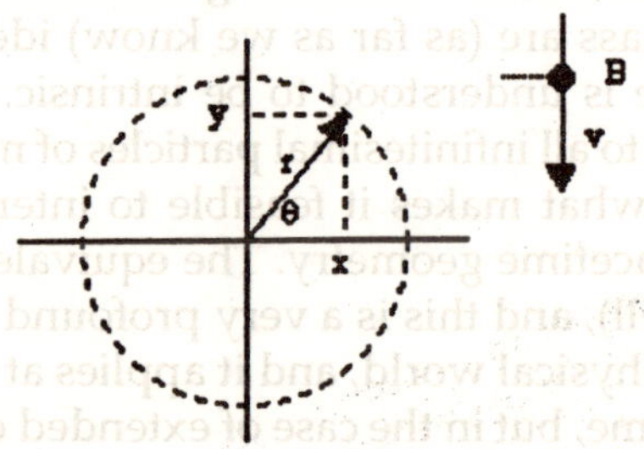

With respect to the Cartesian coordinates we have the familiar Pythagorean line element $(ds)^2 = (dx)^2 + (dy)^2$. Also, we know the polar coordinates are related to the Cartesian coordinates by the equations $x = r\cos(\theta)$ and $y = r\sin(\theta)$, so we can evaluate the differentials

$$dx = -r\sin(\theta)d\theta + \cos(\theta)dr \quad dy = r\cos(\theta)d\theta + \sin(\theta)dr$$

which of course are the transformation equations for the covariant metric tensor. Substituting these differentials into the Pythagorean metric equation, we have the metric for polar coordinates $(ds)^2 = (dr)^2 + r^2 (d\theta)^2$. Therefore, the covariant and contravariant metric tensors for these polar coordinates are

$$g_{\mu\nu} = \begin{bmatrix} 1 & 0 \\ 0 & r^2 \end{bmatrix}$$

and we have the determinant $g = r^2$. The only non-zero partial derivatives of the covariant metric components are $\partial g_{\theta\theta/\partial r=2r}$ and $\partial^2 g_{\theta\theta/\partial r^2=2}$, so the only non-zero Christoffel symbols are $\Gamma^{r}_{\theta\theta} = -r$ and $G^{\theta}_{\theta r} = G^{\theta}_{r\theta} = 1/r$. Inserting these values into (1) gives the geodesic equations for this surface

$$\frac{d^2 r}{ds^2} - r\left(\frac{d\theta}{ds}\right)^2 = 0 \qquad\qquad \frac{d^2 r}{ds^2} + \frac{2}{r}\left(\frac{dr}{ds}\right)\left(\frac{dr}{ds}\right) = 0$$

Since we know this surface is a flat plane, the geodesic curves must be simply straight lines, and indeed it's clear from these equations that any purely radial path (for which $d\theta/ds = 0$) is a geodesic. However, paths going "straight" in the θ direction (at constant r) are not geodesics, and these equations describe how the coordinates must vary along any given trajectory in order to maintain a geodesic path on the plane. Of course, if we insert these polar metric components into Gauss's curvature formula we get K = 0, consistent with the fact that the surface is flat. The reason the geodesics on this surface are not simple linear functions of the coordinates is not because the geodesics are curved, but because the coordinates are curved.

However, if some "flatlanders" living on a small region of this plane were under the impression that the constant-r and constant-q loci were straight, then they might well imagine that the geodesic paths were curved, and that objects which naturally follow those paths are being influenced by some "force field". This may seem far fetched, but suppose we are standing in an upwardly accelerating elevator. We will tend to view things with respect to a coordinate system co-moving with the elevator, and as a result we notice that the natural paths of things are different than they would normally be, as if those objects were being influenced by an additional force field. This is exactly analogous to the flatlanders, except that it is their q axis which is non-linear, whereas our elevator's t axis is non-linear. Inside the accelerating elevator the additional tendency for geodesic paths to "veer off" is not really due to any extra non-linearity of the geodesics, it's due to the non-linearity of the elevator's coordinate system. Hence most people today would say that non-zero Christoffel symbols, by themselves, should not be regarded as indicative of the presence of "true" gravity. If the intrinsic curvature is zero, then non-vanishing Christoffel symbols simply represent the necessary compensation for non-linear coordinates, so, at most (the argument goes) they represent "pseudo-gravity" in such circumstances.

Despite all this Einstein had a valid point that hasn't been fully appreciated by all the subsequent writers of relativity text books. In a letter to his friend Max von Laue in 1950 he tried to explain:

> ...what characterizes the existence of a gravitational field from the empirical standpoint is the non-vanishing of the G^{l}_{ik}, not the non-vanishing of the R_{iklm}. If one does not think intuitively in such a way, one cannot grasp why something like a curvature should have anything at all to do with gravitation. In any case, no reasonable person would have hit upon such a thing. The key for the understanding of the equality of inertial and gravitational mass is missing.

The point of the equivalence principle is that, curving coordinates are gravitation, and there is no intrinsic ontological difference between gravity and pseudo-gravity. On a purely local (infinitesimal) basis, the phenomena of gravity and acceleration were, in Einstein's view, quite analogous to the electric and magnetic fields in the context of special relativity, i.e., they are two ways of looking at (or interpreting) the same thing, in terms of different coordinate systems. Now, it can be argued that there are clear physical differences between electricity and magnetism (e.g., no magnetic monopoles) and how they are "produced" by elementary particle "sources", but one of the keys to the success of special relativity was that it unified the electric and magnetic fields in free space without getting bogged down (as Lorentz did) in trying to fathom the ultimate constituency of elementary charged particles, etc.

Likewise, the most compelling aspect of general relativity is the "left-hand side of the equation" $R_{\mu\nu} - (1/2)g_{\mu\nu}R = T_{\mu\nu}$, which unifies gravity and non-linear coordinates - including acceleration and polar coordinates - in free space, without getting bogged down in the "source" side of the equation, i.e., the fundamental nature of how gravity is ultimately "produced", why the elementary massive particles have the masses they have, and so on. What Einstein was describing to von Laue was the conceptual necessity of identifying the purely geometrical effects of non-inertial coordinates with the physical phenomenon of gravitation. In contrast, the importance and conceptual significance of the curvature (as opposed to the connection) is mainly due to the fact that it defines the mode of coupling of the coordinates with the "source" side of the equation.

Nevertheless, some authors make a point of saying that "gravitational effects are not equivalent to the effects arising from an observer's acceleration...", and they go on to describe how the effects of rotation (for example) can easily be distinguished from typical gravitation produced by massive bodies. In one sense this is obviously true, but on a deeper level it confuses intrinsic and extrinsic qualities by invoking two different limits when imposing the locality condition. Essentially these authors are correctly noting that what is a "sufficiently small" (i.e., local) region of spacetime for transforming away the translatory motion of an object to some degree of approximation may not be sufficiently small for transforming away the rotational motion to the same degree of accuracy, but this does not conflict with the equivalence principle; it just means that for an infinitesimal particle in a rotating body the "sufficiently small" region of spacetime is generally much smaller than for a particle in a non-rotating body, because it must be limited to a small arc of angular travel. Likewise some authors argue that the existence of tidal effects and geodesic deviation conflict with the equivalence principle, but that too is clearly not the case, provided we properly restrict our attention to a sufficiently small region of space and time.

Fundamentally, those who argue against the validity or meaningfulness of the equivalence principle are focusing on the wrong level of attributes when they point out that gravity and acceleration are generally distinguishable. In making this distinction they focus on one or more extrinsic attribute, whereas the equivalence principle is really an assertion of intrinsic equivalence. Indeed, it is the perfect intrinsic equivalence that makes a purely geometrical interpretation of gravity possible. The meaning of the equivalence principle is that the spacetime in a region with a gravitational field is intrinsically the same "stuff" at each point - up to a change of coordinates - as the spacetime in a region without a gravitational field. There is no additional coupling present to produce the effects of gravity on a test body. Gravity is geometry. This is expressed somewhat informally by saying that if we take sufficiently small pieces of curved and flat spacetime we can't tell one from the other, because they are the same stuff. What Einstein called "the happiest thought of my life" was his realization that the perfect equivalence between gravitational and inertial mass noted by Galileo implies that kinematic acceleration and the acceleration of gravity are intrinsically identical, and that this makes possible a purely geometrical interpretation of gravity.

SCHOLIUM

Considering that the first Scholium of Newton's Principia begins with the famous assertion "absolute, true, and mathematical time...flows equably, without relation to anything external", it's ironic that Newton's theory of universal gravitation can be interpreted as a theory of variations in the flow of time. Suppose in Newton's absolute space we establish the Cartesian coordinates x,y,z, and then assign a fourth coordinate, t, to every point. We will call this the coordinate time parameter, but we don't necessarily identify this with the "true time" of events. Instead we postulate that the true lapse of time along an incremental timelike path is dt, given by

$$(g\tau)^2 = g_{00}(dt)^2 + k\left[(dx)^2 + (dy)^2 + (dz)^2\right]$$

From the Galilean standpoint, we assume that a single set of assignments of the time coordinate t to events corresponds to the lapses of proper time dt along any and all paths, which implies that $g_{00} = 1$ and $k = 0$. However, this can only be known to within some observational tolerance. Strictly speaking we can say only that g_{00} is extremely close to 1, and the constant k is very close to zero (in conventional units of measure).

Using indices with $x^0 = t$, $x^1 = x$, $x^2 = y$, and $x^3 = z$, we can re-write (1) as the summation

$$(d\tau)^2 = g_{\mu\nu} dx^\mu dx^\nu$$

where

$$g_{\mu\nu} = \begin{bmatrix} g_{00} & 0 & 0 & 0 \\ 0 & k & 0 & 0 \\ 0 & 0 & k & 0 \\ 0 & 0 & 0 & k \end{bmatrix}$$

Now let's define a four-dimensional array of number representing the second partial derivatives of the g_{bd} as a function of every pair of coordinates x^a, x^c

$$R^a{}_{bcd} = \frac{\partial^2 g_{bd}}{\partial X^a \partial X^c}$$

Also, we define the "contraction" of this array (using the summation convention for repeated indices) as

$$R_{ab} = R^s{}_{asb}$$

Since the only non-zero components of $R^a{}_{bcd}$ are $R^0{}_{0cd}$, it follows that the only non-zero component of R_{ab} is

$$R_{00} = \frac{\partial^2 g_{00}}{\partial t^2} + \frac{\partial^2 g_{00}}{\partial x^2} + \frac{\partial^2 g_{00}}{\partial y^2} + \frac{\partial^2 g_{00}}{\partial z^2}$$

If we assume g_{00} is independent of the coordinate t (meaning that the metrical configuration is static), the first term vanishes and we find that R_{00} is just the Laplacian of g_{00}. Hence if we take our vacuum field equations to be $R_{\mu\nu} = 0$, this is equivalent to requiring that the Laplacian of g_{00} vanish, i.e.,

$$\frac{\partial^2 g_{00}}{\partial x^2}+\frac{\partial^2 g_{00}}{\partial y^2}+\frac{\partial^2 g_{00}}{\partial z^2}=0$$

For convenience let us define the scalar $\varphi = g_{00}/2$. If we consider just spherically symmetrical fields about the origin, we have $\varphi = \varphi$ (r) and so

$$\frac{\partial \varphi}{\partial x}=\frac{d\varphi}{dr}\frac{\partial r}{\partial x} \qquad \frac{\partial^2 \varphi}{\partial x^2}=\frac{d\varphi}{dr}\frac{\partial^2 r}{\partial x^2}+\frac{d^2\varphi}{dr^2}\left(\frac{\partial r}{\partial x}\right)^2$$

and similarly for the partials with respect to y and z. Since $(d\tau)^2 = g_{\mu\nu}dx^\mu dx^\nu$ we have

$$\frac{\partial r}{\partial x}=\frac{x}{r} \qquad \frac{\partial^2 r}{\partial x^2}=\frac{y^2+z^2}{r^3}$$

and similarly for the y and z partials. Making these substitutions back into the Laplace equation gives

$$2\frac{d\varphi}{dr}+r\frac{d^2\varphi}{dr^2}=0$$

This is simple linear differential equation has the unique solution $d\varphi/dr = J/r^2$ where J is a constant of integration, and so we have $\varphi = -J/r + K$ for some constants J and K.

Incidentally, it's worth noting that this applies only in three dimensions. If we were working in just two dimensions, the constant "2" in the above equation would be "1", and the unique solution would be $d\varphi/dr = J/r$, giving $\varphi = J\ln(r) + K$. This shows that Newtonian gravity "works" only with three space dimensions, just as general relativity works only with four spacetime dimensions.

Now that we've solved for the g_{00} field we need the equations of motion. We assume that objects in gravitational free-fall follow geodesics through the spacetime, so the equations of motion are just the geodesic equations

$$\frac{d^2 x^\alpha}{d\tau^2}=-\Gamma^\alpha_{\ \mu\nu}\frac{dx^\mu}{d\tau}\frac{dx^\nu}{d\tau}$$

Where x^α denote the quasi-Euclidean coordinates t,x,y,z defined above. Since we have assumed that the scale factor k between spatial and temporal coordinates is virtually zero, and that g_{00} is nearly equal to unity, it's clear that all the speed components $dx/d\tau$, $dy/d\tau$, $dz/d\tau$ are extremely small, whereas the derivative $dt/d\tau$ is virtually equal to 1. Neglecting all terms containing one or more of the speed components, we're left with the zeroth-order approximation for the spatial accelerations

$$\frac{d^2x}{dt^2} = -\Gamma^x{}_{tt} \qquad \frac{d^2y}{dt^2} = -\Gamma^y{}_{tt} \qquad \frac{d^2z}{dt^2} = -\Gamma^y{}_{tt}$$

From the definition of the Christoffel symbols we have

$$\Gamma^x{}_{tt} = \frac{1}{2}g^{x\alpha}\left[\frac{\partial g_{\alpha t}}{\partial t} + \frac{\partial g_{\alpha t}}{\partial t} - \frac{\partial g_{tt}}{\partial x^\alpha}\right]$$

and similarly for the Christoffel symbols in the y and z equations. Since the metric components are independent of time, the partials with respect to t are all zero. Also, the metric tensor $g_{\mu\nu}$ and its inverse $g^{\mu\nu}$ are both diagonal and the non-zero components of the latter are virtually equal to 1, 1/k, 1/ k, 1/k. All the mixed components of $g^{\mu\nu}$ vanish, so we are left with

$$\Gamma^x{}_{tt} = -\frac{1}{2k}\frac{\partial g_{tt}}{\partial x}$$

and similarly for $\Gamma^y{}_{tt}$ and $\Gamma^z{}_{tt}$. As a result, the equations of motion in the weak slow limit are closely approximated by

$$\frac{d^2x}{dt^2} = \frac{1}{2k}\frac{\partial g_{tt}}{\partial x} \qquad \frac{d^2y}{dt^2} = \frac{1}{2k}\frac{\partial g_{tt}}{\partial y} \qquad \frac{d^2z}{dt^2} = \frac{1}{2k}\frac{\partial g_{tt}}{\partial z}$$

We've seen that the Laplace equation requires g_{tt} to be of the form 2K - 2J/r for some constants K and J in a spherically symmetrical field, and since we expect $dt/d\tau$ to approach 1 as r increases, we can set 2K = 1. With g_{tt} = 1 - 2J/r we have

$$\frac{\partial g_{tt}}{\partial x} = \frac{2J}{r^2}\frac{x}{r}$$

And similarly for the partials with respect to y and z. Therefore the approximate equations of motion in the weak slow limit are

$$\frac{d^2x}{dt^2}=\frac{J/k}{r^2}\left(\frac{x}{r}\right) \qquad \frac{d^2y}{dt^2}=\frac{J/k}{r^2}\left(\frac{y}{r}\right) \qquad \frac{d^2z}{dt^2}=\frac{J/k}{r^2}\left(\frac{z}{r}\right)$$

If we set J/k = -m, i.e., to the negative of the mass of the gravitating source, these are exactly the equations of motion for Newton's inverse-square attraction. Interestingly, this implies that precisely one of J,k is negative. If we choose to make J negative, then the gravitational "potential" has the form $g_{tt} = 1 + 2|J|/r$, which signifies that the potential would increase as we approach the source, as would the rate of proper time along a stationary worldline with respect to coordinate time. In such a universe the value of k would need to be positive in order for gravity to be attractive, i.e., in order for geodesics to converge on the gravitating source. On the other hand, if we choose to make J positive, so that the potential and the rate of proper time decrease as we approach the source, then the constant k must be negative. Referring back to the original line element, this implies an indefinite metric. Naturally we can scale our units so that |k| = 1, but the sign of k is significant. Thus from the observation that "things fall down" we can nearly infer the Minkowski metrical structure of spacetime.

The fact that we can derive the correct trajectories of free-falling objects based on either of two diametrically opposed assumptions is not without precedent. This is very closely related to how Descartes and Newton were able to deduce the correct law of refraction based on the assumption that light travels more rapidly in denser media, while Fermat deduced the same law from the opposite assumption.

In any case, taking k = -1 and J = m, we see that Newton's law of gravitation in the vacuum is $R_{\mu\nu} = 0$, closely paralleling the vacuum field equations of general relativity, which represents the vanishing of the Laplacian of $g_{00}/2$. At a point with non-zero mass density we simply set this equal to -4pr to give Poisson's equation. Hence is we define the energy-momentum array

$$T_{ab} = \begin{bmatrix} \rho & 0 & 0 & 0 \\ 0 & 0 & 0 & 0 \\ 0 & 0 & 0 & 0 \\ 0 & 0 & 0 & 0 \end{bmatrix}$$

we can express Newton's geometrical spacetime law of gravitation as

$$R_{ab} = 8\pi T_{ab}$$

This can be compared with Einstein's field equations

$$R_{ab} = 8\pi\left(T_{ab} - \frac{1}{2}g_{ab}T\right)$$

Of course the "R" and "T" arrays in Newton's law are based on simple partial derivatives, rather than covariant differentiation, so they are not precisely identical to the Ricci tensor and the energy-momentum tensor of general relativity. However, the definitions are close enough that the tensors of general relativity can rightly be viewed as the natural generalizations of the simple Newtonian arrays. The above equations show that the acceleration of gravity is proportional to the rate of change of g_{tt} as a function of r. At any given r we have dt/dt = $\sqrt{g_{tt}}$, so g_{tt} corresponds to the squared "rate of proper time" (with respect to coordinate time) at the given r. It follows that our feet are younger than our heads, because time advances more slowly as we get closer to the centre of the field. So, despite Newton's conception of the perfectly equable flow of time, his theory of gravitation can well be interpreted as a description of the effects of the inequable flow of time. In essence, the effect of Newtonian gravity can be explained in terms of the flow of time being slower near massive objects, and just as a refracted ray of light veers toward the medium in which light goes more slowly (and as a tank veers in the direction of the slower tread-track), objects progressing in time veer in the direction of slower proper time, causing them to accelerate toward massive objects.

GAUSS'S LAW

One of the most famous stories about Gauss depicts him measuring the angles of the great triangle formed by the mountain peaks of Hohenhagen, Inselberg, and Brocken for evidence that the geometry of space is non-Euclidean. It's certainly true that Gauss acquired geodetic survey data during his ten-year involvement in mapping the Kingdom of Hanover during the years from 1818 to 1832, and this data included some large "test triangles", notably the one connecting the those three mountain peaks, which could be used to check for accumulated errors in the smaller triangles. It's also true that Gauss understood how the intrinsic curvature of the Earth's surface would theoretically result in slight discrepancies when fitting the smaller triangles inside the larger triangles, although in practice this effect is negligible, because the Earth's curvature is so slight relative to even the largest triangles that can be visually measured on the surface. Still, Gauss computed the magnitude of this effect for the large test triangles because, as he wrote to Olbers, "the honour of science demands that one understand the nature of this inequality clearly". (The government officials who commissioned Gauss to perform the survey might have recalled Napoleon's remark that Laplace as head of the Department of the Interior had "brought the theory of the infinitely small to administration".) It is sometimes said that the "inequality" which Gauss had in mind was the possible curvature of space itself, but taken in context it seems he was referring to the curvature of the Earth's surface.

On the other hand, if the curvature of space was actually great enough to be observed in optical triangles of this size, then presumably Gauss would have noticed it, so we may still credit him with having performed an empirical observation of geometry, but in this same sense every person who ever lived has made such observations. It might be more meaningful to name people who have explicitly argued against the empirical status of geometry, i.e., who have claimed that the character of spatial relations could be known without empirical observation. In his "Critique of Pure Reason", Kant famously declared that Euclidean geometry is the only possible way in which the mind can organize information about extrinsic spatial relations. One could also cite Plato and other idealists and a priorists. On the other hand, Poincare advocated a conventionalist view of geometry, arguing that we can always, if we wish, cast our physics within a Euclidean spatial framework - provided we are prepared to make whatever adjustments in our physical laws are necessary to preserve this convention. In any case, it seems reasonable to agree with Buhler, who concludes in his biography of Gauss that "the oft-told story according to which Gauss wanted to decide the question [of whether space is perfectly Euclidean] by measuring a particularly large triangle is, as far as we know, a myth."

The first person to publicly propose an actual test of the geometry of space was apparently Lobachevski, who suggested that one might "investigate a stellar triangle for an experimental resolution of the question." The "stellar triangle" he proposed was the star Sirius and two different positions of the Earth at 6-month intervals. This was used by Lobachevski as an example to show how we could place limits on the deviation from flatness of actual space, based on the fact that, in a hyperbolic space of constant curvature, there is a limit to how small a star's parallax can be, even the most distant star. Gauss had already (in private correspondence with Taurinus in 1824) defined the "characteristic length" of a hyperbolic space, which he called "k", and had derived several formulas for the properties of such a space in terms of this parameter. For example, the circumference of a circle of radius r in a hyperbolic space whose "characteristic length" is k is given by

$$C = \pi k\left(e^{r/k} - e^{-r/k}\right) = 2\pi k \sinh\left(\frac{r}{k}\right)$$

Since $\sinh(x) = x + x^3/3! + \ldots$, it follows that C approaches $2\pi r$ as k increases to infinity. From the fact that the maximum parallax of Sirius (as seen from the Earth at various times) is 1",24, Lobachevski deduced that the value of k for our space must be at least 166,000 times the radius of the Earth's orbit. Naturally the same analysis for more distant stars gives an even larger lower bound on k.

The first definite measurement of parallax for a fixed star was performed by Friedrich Bessel (a close friend of Gauss) in 1838, on the star 61 Cygni. Shortly thereafter he measured Sirius (and discovered its binary nature). Lobachevski's first paper on "the new geometry" was presented as a lecture at Kasan in 1826, followed by publications in 1829, 1835, 1840, and 1855 (a year before his death). He presented his lower bound on "k" in the later editions based on the still fairly recent experimental results of stellar parallax measurements. In 1855 Lobachevski was completely blind, so he dictated his exposition.

The other person credited with discovering non-Euclidean geometry, Janos Bolyai, was the son of Wolfgang Bolyai, who was a friend (almost the only friend) of Gauss during their school days at Gottingen in the late 1790's. The elder Bolyai had also been interested in the foundations of geometry, and spent many years trying to prove that Euclid's parallel postulate is a consequence of the other postulates. Eventually he concluded that it had been a waste of time, and he became worried when his son Janos became interested in the same subject. The alarmed father wrote to his son

> For God's sake, I beseech you, give it up. Fear it no less than sensual passions because it, too, may take all your time, and deprive you of your health, peace of mind, and happiness in life.

Undeterred, Janos continued to devote himself to the study of the parallel postulate, and in 1829 he succeeded in proving just the opposite of what his father (and so many others) had tried in vain to prove. Janos found (as had Gauss, Taurinnus, and Lobachevesky just a few years earlier) that

Euclid's parallel postulate is not a consequence of the other postulates, but is rather an independent assumption, and that alternative but equally consistent geometries based on different assumptions may be constructed. He called this the "Absolute Science of Space", and wrote to his father that "I have created a new universe from nothing". The father then, forgetting his earlier warnings, urged Janos to publish his findings as soon as possible, noting that

> ...ideas pass easily from one to another, and secondly... many things have an epoch, in which they are found at the same time in several places, just as violets appear on every side in spring.

Naturally the elder Bolyai sent a copy of his son's spectacular discovery to Gauss, in June of 1831, but it was apparently lost in the mail. Another copy was sent in January of 1832, and then seven weeks later Gauss sent a reply to his old friend:

> If I commenced by saying that I am unable to praise this work, you would certainly be surprised for a moment. But I cannot say otherwise. To praise it would be to praise myself. Indeed the whole contents of the work, the path taken by your son, the results to which he is led, coincide almost entirely with my meditations, which have occupied my mind partly for the last thirty or thirty-five years. So I remained quite stupefied. So far as my own work is concerned, of which up till now I have put little on paper, my intention was not to let it be published during my lifetime. ... I have found very few people who could regard with any special interest what I communicated to them on this subject. ...it was my idea to write down all this later so that at least it should not perish with me. It is therefore a pleasant surprise for me that I am spared this trouble, and I am very glad that it is just the son of my old friend, who takes the precedence of me in such a remarkable manner.

In his later years Gauss' response to many communications of new mathematical results was similar to the above. For example, he once remarked that a paper of Abel's saved him the trouble of having to publish about a third of his results concerning elliptic integrals. Likewise he confided to friends that Jacobi and Eisenstein had "spared him the trouble" of publishing important results that he (Gauss) had possessed since he was a teenager, but had never bothered to publish. Dedekind even reports that Gauss made a similar comment about Riemann's dissertation. It's true that Gauss' personal letters and notebooks substantiate to some extent his private claims of priority for nearly every major mathematical advance of the 19th century, but the full extent of his early and unpublished accomplishments did not become known until after his death, and in any case it wouldn't have softened the blow to his contemporaries. Janos Bolyai was so embittered by Gauss's backhanded response to his non-Euclidean geometry that he never published again.

As another example of what Wolfgang Bolyai called "violets appearing on every side", Maxwell's great 1865 triumph of showing that electromagnetic waves propagate at the speed of light was, to some degree, anticipated by others. In 1848 Kirchoff had noted that the ratio of electromagnetic and electrostatic units was equal to the speed of light, although he gave no explanation for this coincidence. In 1858 Riemann presented a theory based on the hypothesis that electromagnetic effects propagate at a fixed speed, and then deduced that this speed must equal the ratio of electromagnetic and electrostatic units, i.e., $1/\sqrt{\varepsilon_0\mu_0}$.

Even in this field we find that Gauss can plausibly claim priority for some interesting developments. Recall that, in addition to being the foremost mathematician of his day, Gauss was also prominent in studying the phenomena of electricity and magnetism (in fact the unit of magnetism is called a Gauss), and even dabbled in electrodynamics. He reached the conclusion that the keystone of electrodynamics would turn out to depend on an understanding of how electric effects propagate in time. In 1835 he wrote (in an unpublished papers, discovered after his death) that

> Two elements of electricity in a state of relative motion attract or repel one another, but not in the same way as if they are in a state of relative rest.

He even suggested the following mathematical form for the complete electromagnetic force F between two particles with charges q_1 and q_2 in arbitrary states of motion

$$F = \frac{q_1 q_2}{r^2}\left(1 + \frac{1}{c^2}\left(u^2 - \frac{3}{2}\dot{r}^2\right)\right) \qquad F = \frac{q_1 q_2}{r^2}\left(1 + \frac{1}{c^2}\left(r\ddot{r} - r.\ddot{r} - \frac{1}{2}\dot{r}^2\right)\right)$$

where r is the scalar distance, r is the vector distance, u is the relative velocity between the particles, and dots signify derivatives with respect to time. This formula actually gives the correct results for particles in uniform (inertial) motion, in which case the second derivative of the vector r is zero. However, the dot product in Gauss's formula violates conservation of energy for general motions. A few years later (in 1845), Gauss's friend Wilhelm Weber proposed a force law identical to Gauss's, except he excluded the dot product, i.e., he proposed the formula

$$F = \frac{q_1 q_2}{r^2}\left(1 + \frac{1}{2}\left(\frac{\dot{r}}{c}\right)^2 \frac{r\ddot{r}}{c^2}\right)$$

Weber pointed out that, unlike Gauss's original formula, this force law satisfies conservation of energy, as shown by the fact that it can be derived from the potential function

$$\psi = \frac{q_1 q_2}{r}\left(1 - \frac{1}{2}\left(\frac{\dot{r}}{2}\right)^2\right)$$

In terms of this potential, the force given by F = -dψ/dr is precisely Weber's force law. This equation was used by Weber as the basis of his theory of electrodynamics published in 1846. Indeed this formula served as the basis for most theoretical studies of electromagnetism until it was finally superseded by Maxwell's theory beginning in the 1870s. It's interesting that in order for energy to be conserved it was necessary to eliminate the vectors from Gauss's formula, making the result entirely in terms of the scalar distance and its derivatives. Compare this with the separation equations discussed in sections above . Note that according to equation above the condition for the force between two charged particles to vanish is that the quantity in parentheses equals zero, i.e.,

$$\frac{1}{2}\dot{r}^2 - r\ddot{r} = c^2$$

Differentiating both sides and dividing by r gives the condition $\dddot{r} = 0$, if we set N = 0. (The vanishing of the third derivative is also the condition for zero radiation reaction according to the Lorentz-Dirac equations of classical electrodynamics.) Interestingly, Kurt Schwarzschild published a paper in 1903 describing in detail how the Gauss-Weber approach could actually have been developed into a viable theory. In any case, if the two charged particles are separating (without rotation) at a uniform speed $\dot{r} = v$, Gauss' formula relates the electrostatic force $F_0 = q_1q_1/r^2$ to the dynamic force as

$$\frac{F}{F_0} = 1 - \frac{1}{2}\left(\frac{v}{c}\right)^2 \approx \sqrt{1 - \left(\frac{v}{c}\right)^2}$$

So, to press the point, one could argue that Gauss' offhand suggestion for the formula expressing electrodynamic force already represents the seeds of Lorentz's molecular force hypothesis, from which follows the length contraction and time dilation of the Lorentz transformations and special relativity. In fact, pursuing this line of thought, Riemann (one of Gauss' successors at Gottingen) proposed in 1858 that the electric potential should satisfy the equation

$$\frac{\partial^2\psi}{\partial x^2} - \frac{\partial^2\psi}{\partial y^2} + \frac{\partial^2\psi}{\partial z^2} - \frac{1}{c^2}\frac{\partial^2\psi}{\partial t^2} = -4\pi\rho$$

Where ρ is the charge density. This equation does indeed give the retarded electrostatic potential, which, combined with the similar equation for the vector potential, serves as the basis for the whole classical theory of electromagnetism. Assuming conservation of charge, the invariance of the Minkowski spacetime metric clearly emerges from this equation, as does the invariance of the speed of light in terms of any suitable (i.e., inertially homogeneous and isotropic) system coordinates.

Chapter 7

Theory of Relativity

POSTULATES

As discussed in previous sections, even after stipulating the existence of coordinate systems with respect to which inertia is homogeneous and isotropic, there remains a fundamental amgibuity as to the character of the relationship between relatively moving inertial coordinate systems, corresponding to three classes of possible metrical structures, with the k values -1, 0, and +1. There is a remarkably close historical analogy for this situation, dating back to one of the first formal systems of thought ever proposed. In Book I of The Elements, Euclid consolidated and systematized plane geometry as it was known circa 300 BC into a formal deductive system. As it has come down to us, it is based on five postulates together with several definitions and common notions. (It's worth noting, however, that the classifications of these premises was revised many times in various translations.) The first four of these postulates are stated very succinctly

1. A straight line may be drawn from any point to any other point.
2. A straight line segment can be uniquely and indefinitely extended.
3. We may draw a circle of any radius about any point.
4. All right angles are equal to one another.

Each of these assertions actually entails a fairly complicated set of premises and ambiguities, but they were accepted as unobjectionable for two thousand years. However, Euclid's final postulate was regarded with suspicion from earliest times. It has a very different appearance from the others - a difference that neither Euclid nor his subsequent editors and translators attempted to disguise. The fifth postulate is expressed as follows:

5. If a straight line falling on two straight lines makes the [sum of the] interior angles on the same side less than two right angles, then the two straight lines, if produced indefinitely, meet on that side on which the angles are less than two right angles.

This postulate is equivalent to the statement that there's exactly one line through a given point P parallel to a given line L, as illustrated below-

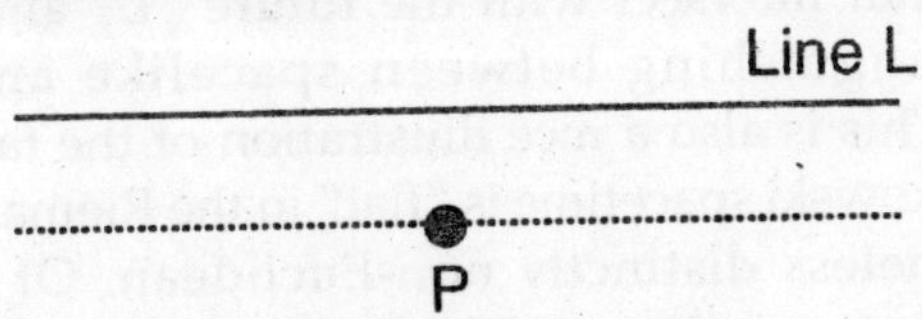

Although this proposition is fairly plausible (albeit somewhat awkward to state), many people suspected that it might be logically deducible from the other postulates, axioms, and common notions. There were also attempts to substitute for Euclid's fifth postulate a simpler or more self-evident proposition. However, we now understand that Euclid's fifth postulate is logically independent of the rest of Euclid's logical structure. In fact, it's possible to develop logically consistent geometries in which Euclid's fifth postulate is false. For example, we can assume that there are infinitely many lines through P that are parallel to (i.e., never intersect) the line L. It might seem (at first) that it would be impossible to reason with such an assumption, that it would either lead to contradictions or else cause the system to degenerate into a logical triviality about which nothing interesting could be said, but, remarkably, this turns out not to be the case.

Suppose that although there are infinitely many lines through P that never intersect L, there are also infinitely many that do intersect L. This, combined with the other axioms and postulates of plane geometry, implies that there are two lines through P defining the boundary between lines that do intersect L and lines that don't, as shown below:

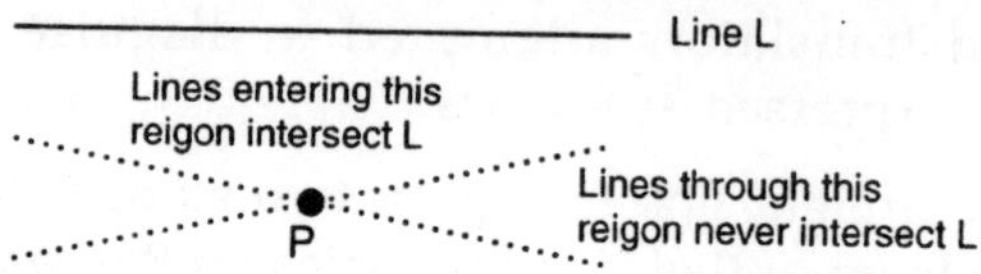

This leads to the original non-Euclidean geometry of Lobachevski, Bolyai, and Gauss, i.e., the hyperbolic plane. The analogy to Minkowski spacetime is obvious. The behaviour of "straight lines" in a surface of negative curvature (although positive-definite) is nicely suggestive of how the light-lines in spacetime serve as the dividing lines between those lines through P that intersect with the future "L" and those that don't (distinguishing between spacelike and timelike intervals). This is also a nice illustration of the fact that even though Minkowski spacetime is "flat" in the Riemannian sense, it is nevertheless distinctly non-Euclidean. Of course, the possibility that spacetime might be curved as well as locally Minkowskian led to general relativity, but arguably the conceptual leap required to go from a positive-definite to a non-positive-definite metric is greater than that required to go from a flat to a curved metric. The former implies that the local geometrical structure of the effective spatio-temporal manifold of events is profoundly different than had been assumed for thousands of years, and this realization led naturally to a new set of principles with which to organize and interpret our experience.

It became clear in the 19th century that there are actually three classes of geometries consistent with Euclid's basic premises, depending on what we adopt as the "fifth postulate". The three types of geometry correspond to spaces of negative, positive, or zero curvature. The analogy to the three possible classes of spacetimes (Euclidean, Galilean, and Minkowskian) is obvious, and in both cases it came to be recognized that, insofar as these mathematical structures were supposed to represent physical properties, the choice between the alternatives was a matter for empirical investigation.

Nevertheless, the superficially axiomatic way in which Einstein presented the special theory in his 1905 paper tended to encourage the idea that special relativity represented a closed formal system, like Euclid's geometry interpreted in the purely mathematical sense. For example, in 1907 Paul Ehrenfest wrote that

> In the formulation in which Mr Einstein published it, Lorentzian relativistic electrodynamics is rather generally viewed as a complete system. Accordingly it must be able to provide an answer purely deductively to the question [involving the shape of the moving electron]...

However, Einstein himself was quick to disavow this idea, answering

> The principle of relativity, or, more exactly, the principle of relativity together with the principle of the constancy of the velocity of light, is not to be conceived as a "complete system," in fact, not as a system at all, but merely as a heuristic principle which, when considered by itself, contains only statements about rigid bodies, clocks, and light signals. It is only by requiring relations between otherwise seemingly unrelated laws that the theory of relativity provides additional statements.

Just as the basic premises of Euclid's geometry were classified in many different ways (e.g., postulates, axioms, common notions, definitions), the premises on which Einstein based special relativity can be classified in many different ways. Indeed, in his 1905 paper, Einstein introduced the first of these premises as follows:

> ... the same laws of electrodynamics and optics will be valid for all coordinate systems in which the

> equations of mechanics hold good. We will raise this conjecture (hereafter called the "principle of relativity") to the status of a postulate...

Here, in a single sentence, we find a proposition referred to as a conjecture, a principle, and a postulate. The meanings of these three terms are quite distinct, but they are each arguably applicable. The assertion of the co-relativity of optics and mechanics was, and will always be, conjectural, because it can be empirically corroborated only up to a limited precision. Einstein formally adopted this conjecture as a postulate, but on a more fundamental level it serves as a principle, since it entails the decision to organize our knowledge in terms of coordinate systems with respect to which the equations of mechanics hold good, i.e., inertial coordinate systems. Einstein goes on to introduce a second proposition that he formally adopts as a postulate, namely,

> ... that the velocity of light always propagates in empty space with a definite velocity c that is independent of the state of motion of the emitting body. These two postulates suffice for the attainment of a simple and consistent electrodynamics of moving bodies based on Maxwell's theory for bodies at rest.

Interestingly, in the paper "Does the Inertia of a Body Depend on Its Energy Content?" published later in the same year, Einstein commented that

> ... the principle of the constancy of the velocity of light... is of course contained in Maxwell's equations.

In view of this, some have wondered why he did not simply dispense with his "second postulate" and assert that the "laws of electrodynamics and optics" in the statement of the first principle are none other than Maxwell's equations. In other words, why didn't he simply base his theory on the single proposition that Maxwell's equations are valid for every system of coordinates in which the laws of mechanics hold good? Part of the answer is that he realized important parts of physics, such as the physics of elementary particles, cannot possibly be explained in terms of Maxwellian electrodynamics. In a note published in 1907 he wrote

> It should be noted that the laws that govern [the structure of the electron] cannot be derived from electrodynamics alone. After all, this structure necessarily results from the introduction of forces which balance the electrodynamic ones.

More fundamentally, by 1905 he was already aware of the fact that, although Maxwell's equations are empirically satisfactory in many respects, they cannot be regarded as fundamentally correct or valid. In his paper "On a Heuristic Point of View Concerning the Production and Transformation of Light" he wrote

> ... despite the complete confirmation of [Maxwell's theory] by experiment, the theory of light, operating with continuous spatial functions, leads to contradictions when applied to the phenomena of emission and transformation of light.

Thus it isn't surprising that he chose not to base the theory of relativity on Maxwell's equations. He needed to distill from electromagnetic phenomena the key feature whose significance "transcended its connection with Maxwell's equations", and which would serve as a viable principle for organizing our knowledge of all phenomena, including both optics and mechanics. The principle he selected was the existence of an invariant speed with respect to any local system of inertial coordinates, and then for definiteness he could identify this speed with the speed of propagation of electromagnetic energy.

After reviewing the operational definition of inertial coordinates (which he does by optical rather than mechanical means, thereby missing an opportunity to clarify the significance of inertial coordinates in establishing the connection between mechanical and optical phenomena), he gives more formal statements of his two principles

The following reflections are based on the principle of relativity and the principle of the constancy of the velocity of light. These two principles we define as follows:

1. The laws by which the states of physical systems undergo change are not affected, whether these changes of state be referred to the one or the other of two systems of co-ordinates in uniform translatory motion.
2. Any ray of light moves in the "stationary" system of co-ordinates with the determined velocity c, whether the ray is emitted by a stationary or by a moving body. Hence velocity equals [length of] light path divided by time interval [of light path], where time interval [and length are] to be taken in the sense of the definition.

The first of these is nothing but the principle of inertial relativity, which had been accepted as a fundamental principle of physics since the time of Galileo. Strictly speaking, Einstein's statement of the principle here is incorrect, because he assumes the coordinate systems in which the equations of mechanics hold good are fully characterized by being in uniform translatory motion, whereas in fact it is also necessary to specify an inertially isotropic simultaneity. Einstein chose to address this aspect of inertial coordinate systems by means of a separate and seemingly arbitrary definition of simultaneity based on optical phenomena, which unfortunately has invited much misguided philosophical debate about what should be considered "true" simultaneity. All this could have been avoided if, from the start, Einstein had merely stated that an inertial coordinate system is one in which mechanical inertia is homogeneous and isotropic (just as Galileo said), and then noting that automatically entails the conventional choice of simultaneity. The content of his first principle (i.e., the relativity principle) is simply that the inertial simultaneity of mechanics and the optical simultaneity of electrodynamics are identical.

Despite the shortcomings of its statement, the principle of relativity was very familiar to the physicists of 1905, whether they wholeheartedly accepted it or not. Einstein's second principle, by itself, was also not regarded as particularly novel, because it conveys the usual understanding of how a wave propagates at a fixed speed through a medium, independent of the speed of the source. It was the combination of these two principles that was new, since they had previously been thought to be irreconcilable. In a sense, the first principle arose from the "ballistic particles in a vacuum" view of physics, and the second arose from the "wave in a material medium" view of physics. Both of these views can trace their origins back to ancient times, and both seem to capture some fundamental truth about the world, and yet they had always been regarded as mutually exclusive. Einstein's achievement was to explain how they could be reconciled.

Of course, Einstein's second principle it isn't a self-contained statement, because its entire meaning and significance depends on "the sense of" time intervals and (implicitly) spatial lengths given in §1, where we find that time intervals and spatial lengths are defined to be such that their ratio equals the fixed constant c for light paths. This has tempted some readers to conclude that "Einstein's second principle" was merely a tautology, with no substantial content. The source of this confusion is the fact that the essential axiomatic foundations underlying special relativity are contained not in the two famous propositions at the beginning of §2 of Einstein's paper (as quoted above), but rather in the sequence of assumptions and definitions explicitly spelled out in §1. Among these are the very first statement

> Let us take a system of co-ordinates in which the equations of Newtonian mechanics hold good.

In subsequent re-prints of this paper Sommerfeld added a footnote to this statement, to say "i.e., to the first approximation", meaning for motion with speeds small in comparison with the speed of light. (This illustrates the difficulty of writing a paper that results in a modification of the equations of Newtonian mechanics!) Of course, Einstein was aware of the epistemological shortcomings of the above statement, because while it tells us to begin with an inertial system of coordinates, it doesn't tell us how to identify such a system. This has always been a potential source of ambiguity for mechanics based on the principle of inertia. Strictly speaking, Newton's laws are epistemologically circular, so in practice we must apply it both inductively and deductively. First we use them inductively with our primitive observations to identify inertial coordinate systems by observing how things behave. Then at some point when we've gained confidence in the inertialness of our coordinates, we begin to apply the laws deductively, i.e., we begin to deduce how things will behave with respect to our inertial coordinates. Ultimately this is how all physical theories are applied, first inductively as an organizing principle for our observations, and then deductively as "laws" to make predictions. Neither Galilean nor special relativity is able to justify the privileged role given to a particular class of coordinate systems, nor to provide a non-circular means of identifying those systems. In practice we identify inertial systems by means of an incomplete induction. Although Einstein was aware of the deficiency of this approach (which he subsequently labored to eliminate from the general theory), in 1905 he judged it to be the only pragmatic way forward.

The next fundamental assertion in §1 of Einstein's paper is that lengths and time intervals can be measured by (and expressed in terms of) a set of primitive elements called "measuring rods" and "clocks". Again Einstein was fully aware of the deficiency. As he later wrote

> It is striking that the theory introduces (in addition to the four-dimensional space) two kinds of physical things, namely, (1) measuring rods and clocks, and (2) all other things, including the electromagnetic field, the material point, and so on. This, in a certain sense, is inconsistent; strictly speaking, measuring rods and clocks should emerge as solutions of the basic

> equations (objects consisting of moving atomic configurations), not, as it were, as theoretically self-sufficient entities. ...it was clear from the very beginning that the postulates of the theory are not strong enough to deduce from them equations for physical events sufficiently complete and sufficiently free from arbitrariness in order to base upon such a foundation a theory of measuring rods and clocks...
> it was better to admit such inconsistency - with the obligation, however, of eliminating it at a later stage of the theory...

Thus the introduction of clocks and rulers as primitive entities was another pragmatic concession, and one that Einstein realized was not strictly justifiable on any other grounds than provisional expediency.

Next Einstein acknowledges that we could content ourselves to time events by using an observer located at the origin of the coordinate system, which corresponds to the absolute time of Lorentz, as discussed earlier. Following this he describes the "much more practical arrangement" based on the reciprocal operational definition of simultaneity. He says

We assume this definition of synchronization to be free of any possible contradictions, applicable to arbitrarily many points, and that the following relations are universally valid:

1. If the clock at B synchronizes with the clock at A, the clock at A synchronizes with the clock at B.
2. If the clock at A synchronizes with the clock at B and also with the clock at C, the clocks at B and C also synchronize with each other.

These are important and non-trivial assumptions about the viability of the proposed operational procedure for synchronizing clocks, but they are only indirectly invoked by the reference to "the sense of time intervals" in the statement of Einstein's second principle. Furthermore, Einstein himself subsequently identified at least three more assumptions (homogeneity, spatial isotropy, memorylessness) that are tacitly invoked in the formal development of special relativity. The list of unstated assumptions would actually be even longer if we were to construct a theory beginning from nothing but an individual's primitive sense perceptions. The justification for leaving them out of a scientific paper is that these can mostly be classified as what Euclid called "common notions", i.e., axioms that are common to all fields of thought.

In many respects Einstein modeled his presentation of special relativity not on Euclid's Elements (as Newton had done in the Principia), but on the formal theory of thermodynamics, which is founded on the principle of the conservation of energy. There are different kinds of energy, with formally different units, e.g., mechanical and gravitational potential energy are typically measured in terms of joules (a force times a distance, or equivalently a mass times a squared velocity), whereas heat energy is measured in calories (the amount of heat required to raise the temperature of 1 gram of water by one degree C). It's far from obvious that these two things can be treated as different aspects of the same thing, i.e., energy. However, through careful experiments and observations we find that whenever mechanical energy is dissipated by friction (or any other dissipative process), the amount of heat produced is proportional to the amount of mechanical energy dissipated. Conversely, whenever heat is involved in a process that yields mechanical work, the heat content is reduced in proportion to the amount of work produced. In both cases the constant of proportionality is found to be 4.1833 joules per calorie.

Now, the First Law of thermodynamics asserts that the total energy of any physical process is always conserved, provided we "correctly" account for everything. Of course, in order for this assertion to even make sense we need to define the proportionality constants between different kinds of energy, and those constants are naturally defined so as to make the First Law true. In other words, we determine the proportionality between heat and mechanical work by observing these quantities and assuming that those two changes represent equal quantities of something called "energy". But this assumption is essentially equivalent to the First Law, so if we apply these operational definitions and constants of proportionality, the conservation of energy can be regarded as a tautology or a convention.

This shows clearly that, just as in the case of Newton's laws, these propositions are actually principles rather than postulates, meaning that they first serve as organizing principles for our measurements and observations, and only subsequently do they serve as "laws" from which we may deduce further consequences. This is the sense in which fundamental physical principles always operate. Wein's letter of 1912 nominating Einstein and Lorentz for the Nobel prize commented on this same point, saying that "the confirmation of [special relativity] by experiment... resembles the experimental confirmation of the conservation of energy".

Einstein himself acknowledged that he consciously modeled the formal structure of special relativity on thermodynamics. He wrote in his autobiographical notes

> Gradually I despaired of the possibility of discovering the true laws by means of constructive efforts based on known facts. The longer and the more desperately I tried, the more I came to the conviction that only the discovery of a universal formal principle could lead us to assured results. The example I saw before me was thermodynamics. The general principle was there given in the proposition: The laws of nature are such that it is impossible to construct a perpetuum mobile (of the first and second kinds).

This principle is a meta-law, i.e., it does not express a particular law of nature, but rather a general principle to which all the laws of nature conform. In 1907 Ehrenfest suggested that special relativity constituted a closed axiomatic system, but Einstein quickly replied that this was not the case. He explained that the relativity principle combined with the principle of invariant light speed is not a closed system at all, but rather it provides a coherent framework within which to conduct physical investigations. As he put it, the principles of special relativity "permit certain laws to be traced back to one another (like the second law of thermodynamics)."

Not only is there a close formal similarity between the axiomatic structures of thermodynamics and special relativity, each based on two fundamental principles, these two theories are also substantively extensions of each other. The first law of thermodynamics can be placed in correspondence with the basic principle of relativity, which suggests the famous relation $E = mc^2$, thereby enlarging the realm of applicability of the first law. The second law of thermodynamics, like Einstein's second principle of invariant light speed, is more sophisticated and more subtle. A physical process whose net effect is to remove heat from a body and produce an equivalent amount of work is called perpetual motion of the second kind. It isn't obvious from the first law that such a process is impossible, and indeed there were many attempts to find such a process - just as there were attempts to identify the rest frame of the electromagnetic ether - but all such attempts failed. Moreover, they failed in such a way as to make it clear that the failures were not accidental, but that a fundamental principle was involved.

In the case of thermodynamics this was ultimately formulated as the second law, one statement of which (as alluded to by Einstein in the quote above) is simply that perpetual motion of the second kind is impossible - provided the various kinds of energy are defined and measured in the prescribed way. (This theory was Einstein's bread and butter, not only because most of his scientific work prior to 1905 had been in the field of thermodynamics, but also because a patent examiner inevitably is called upon to apply the first and second laws to the analysis of hopeful patent applications.) Compare this with Einstein's second principle, which essentially asserts that it's impossible to measure a speed in excess of the constant c - provided the space and time intervals are defined and measured in the prescribed way. The strength of both principles is due ultimately to the consistency and coherence of the ways in which they propose to analyze the processes of nature.

Needless to say, our physical principles are not arbitrarily selected assumptions, they are hard-won distillations of a wide range of empirical facts. Regarding the justification for the principles on which Einstein based special relativity, many popular accounts give a prominent place to the famous experiments of Michelson and Morley, especially the crucial version performed in 1889, often presenting this as the "brute fact" that precipitated relativity. Why, then, does Einstein's 1905 paper fail to cite this famous experiment? It does mention at one point "the various unsuccessful attempts to measure the Earth's motion with respect to the ether", but never refers to Michelson's results specifically. The conspicuous absence of any reference to this important experimental result has puzzled biographers and historians of science. Clearly Einstein's intent was to present the most persuasive possible case for the relativity of space and time, and Michelson's results would (it seems) have been a very strong piece of evidence in his favour. Could he simply have been unaware of the experiment at the time of writing the paper?

Einstein's own recollections on this point were not entirely consistent. He sometimes said he couldn't remember if he had been aware in 1905 of Michelson's experiments, but at other times he acknowledged that he had known of it from having read the works of Lorentz. Indeed, considering Einstein's obvious familiarity with Lorentz's works, and given all the attention that Lorentz paid to Michelson's ether drift experiments over the years, it's difficult to imagine that Einstein never absorbed any reference to those experiments. Assuming he was aware of Michelson's results prior to 1905, why did he chose not to cite them in support of his second principle? Of course, his paper includes no formal "references" at all (which in itself seems peculiar, especially to modern readers accustomed to extensive citations in scholarly works), but it does refer to some other experiments and theories by name, so an explicit reference to Michelson's result would not have been out of place.

One possible explanation for Einstein's reluctance to cite Michelson, both in 1905 and subsequently, is that he was sophisticated enough to know that his "theory" was technically just a re-interpretation of Lorentz's theory - making identical predictions - so it could not be preferred on the basis of agreement with experiment. To Einstein the most important quality of his interpretation was not its consistency with experiment, but it's inherent philosophical soundness. In other words, conflict with experiment was bad, but agreement with experiment by means of ad hoc assumptions was hardly any better. His critique of Lorentz's theory (or what he knew of it at the time) was not so much that it was empirically "wrong" (which it wasn't), but that the length contraction and time dilation effects had been inserted ad hoc to match the null results Michelson. Therefore, Einstein would naturally have been concerned to avoid giving the impression that his relativistic theory had been contrived specifically to conform with Michelson's results. He may well have realized that any appeal to the Michelson-Morley experiment in order to justify his theory would diminish rather than enhance its persuasiveness.

This is not to suggest that Einstein was being disingenuous, because it's clear that the principles of special relativity actually do emerge very naturally from just the first-order effects of magnetic induction (for example), and even from more basic considerations of the mathematical intelligibility of Galilean versus Lorentzian transformations (as stressed by Minkowski in his famous 1908 lecture). It seems clear that Einstein's explanations for how he arrived at special relativity were sincere expressions of his beliefs about the origins of special relativity in his own mind. He was focused on the phenomenon of magnetic induction and the unphysical asymmetry of the pre-relativistic explanations. This was combined with a strong instinctive belief in the complete relativity of physics. He told Shankland in 1950 that the experimental results which had influenced him the most were stellar aberration and Fizeau's measurements on the speed of light in moving water. "They were enough," he said.

NATURAL AND VIOLENT MOTIONS

The concept of force is one of the most peculiar in all of physics. It is, in one sense, the most viscerally immediate concept in classical mechanics, and seems to serve as the essential "agent of causality" in all interactions, and yet the ontological status of force has always been highly suspect. We sometimes regard force as the cause of changes in motion, and imagine that those changes would not occur in the absence of the forces, but this causative aspect of force is an independent assumption that does not follow from any quantifiable definition, since we could equally well regard force as being caused by changes in motion, or even as merely a descriptive parameter with no independent ontological standing at all.

In addition, there is an inherent ambiguity in the idea of changes in motion, because it isn't obvious what constitutes unchanging (i.e., unforced) motion. Aristotle believed it was necessary to distinguish between two fundamentally distinct kinds of motion, which he called natural motions and violent motions. The natural motions included the apparent movements of celestial objects, the falling of leaves to the ground, the upward movement of flames and hot gases in the atmosphere, or of air bubbles in water, and so on. According to Aristotle, the cause of such motions is that all objects and substances have a natural place or level (such as air above, water below), and they proceed in the most direct way, along straight vertical paths, to their natural places. The motion of the celestial bodies is circular because this is the most perfect kind of unchanging eternal motion, whereas the necessarily transitory motions of sublunary objects are rectilinear. It may not be too misleading to characterize Aristotle's concept of sublunary motion as a theory of buoyancy, since the natural place of light elements is above, and the natural place of heavy elements is below. If an object is out of place, it naturally moves up or down as appropriate to reach its proper place.

Aristotle has often been criticized for saying (or seeming to say) that the speed at which an object falls (through the air) is proportional to its weight. To the modern reader this seems absurd, as it is contradicted by the simplest observations of falling objects. However, it's conceivable that we misinterpret Aristotle's meaning, partly because we're so accustomed to regarding the concept of force as the cause of motion, rather than as an effect or concomitant attribute of motion. If we consider the downward force (which Aristotle would call the weight) of an object to be the force that would be required to keep it at its current height, then the "weight" of an object really is substantially greater the faster it falls. More strength is required to catch a falling object than to hold the same object at rest. Some Aristotelian scholars have speculated that this was Aristotle's actual meaning, although his writing's on the subject are so sketchy that we can't know for certain. In any case, it illustrates that the concept and significance of force in a physical theory is often murky, and it also shows how thoroughly our understanding of physical phenomena is shaped by the distinction between forces (such as gravity) that we consider to be causes of motion, and those (such as impact forces) that we consider to be caused by motion.

Aristotle also held that the speed of motion was not only proportional to the "weight" (whatever that means) but inversely proportional to the resistance of the medium. Thus his proposed law of motion could be expressed roughly as $V = W/R$, and he used this to argue against the possibility of empty space, i.e., regions in which $R = 0$, because the velocity of any object in such a region would be infinite. This doesn't seem like a very compelling argument, since we could easily counter that the putative vacuum would not be the natural place of any object, so it would have no "weight" in that direction either. Nevertheless, perhaps to avoid wrestling with the mysterious fraction $0/0$, Aristotle surrounded the four sublunary elements of Earth, Water, Air, and Fire with a fifth element (quintessence), the lightest of all, called aether. This aether filled the super-lunary region, ensuring that we would never need to divide by zero.

In addition to natural motions, Aristotle also considered violent motions, which were any motions resulting from acts of volition of living beings. Although his writings are somewhat obscure and inconsistent in this area, it seems that he believed such beings were capable of self-motion, i.e., of initiating motion in the first instance, without having been compelled to motion by some external agent. Such self-movers are capable of inducing composite motions in other objects, such as when we skip a stone on the surface of a pond. The stone's motion is compounded of a violent component imparted by our hand, and the natural component of motion compelling it toward its natural place (below the air and water). However, as always, we must be careful not to assume that this motion is to be interpreted as the causative result of the composition of two different kinds of forces. It was, for Aristotle, simply the kinematic composition of two different kinds of motion.

The bifurcation of motion into two fundamentally different types, one for natural motions of non-living objects and another for acts of human volition – and the attention that Aristotle gave to the question of unmoved movers, etc. – is obviously related to the issue of free will, and demonstrates the strong tendency of scientists in all ages to exempt human behaviour from the natural laws of physics, and to regard motions resulting from human actions as original, in the sense that they need not be attributed to other motions. Refer that Aristotle's distinction between natural and violent motions plays a key role in the analysis of certain puzzling aspects of quantum theory.

We can also see that the ontological status of "force" in Aristotle's physics is ambiguous. In some circumstances it seems to be more an attribute of motion rather than a cause of motion. Even if we consider the quantitative physics of Galileo, Newton, and beyond, it remains true that "force", while playing a central role in the formulation, serves mainly as an intermediate quantity in the calculations. In fact, the concept of 'force' could almost be eliminated entirely from classical mechanics. Newton wrestled with the question of whether force should be regarded as an observable or simply a relation between observables. Interestingly, Ernst Mach regarded the third law as Newton's most important contribution to mechanics, even though other's have criticized it as being more a definition than a law.

Newton's struggle to find the "right" axiomatization of mechanics can be seen by reading the preliminary works he wrote leading up to The Principia, such as "De motu corporum in gyrum" (On the motion of bodies in an orbit). At one point he conceived of a system with five Laws of Motion, but what finally appeared in Principia were eight Definitions followed by three Laws. He defined the "quantity of matter" as the measure arising conjointly from the density and the volume. In his critical review of Newtonian mechanics, Mach remarked that this definition is patently circular, noting that "density" is nothing but the quantity of matter per volume. However, all definitions ultimately rely on undefined (irreducible) terms, so perhaps Newton was entitled to take density and volume as two such elements of his axiomatization. Furthermore, by basing the quantity of matter on explicitly finite density and volume, Newton deftly precluded point-like objects with finite quantities of matter, which would imply the existence of infinite forces and infinite potential energy according to his proposed inverse-square law of gravity.

The next basic definition in Principia is of the "quantity of motion", defined as the measure arising conjointly from the velocity and the quantity of matter. Here we see that "velocity" is taken as another irreducible element, like density and volume. Thus, Newton's ontology consists of one irreducible entity, called matter, possessing three primitive attributes, called density, volume, and velocity, and in these terms he defines two secondary attributes, the "quantity of matter" (which we call "mass") as the product of density and volume, and the "quantity of motion" (which we call "momentum") as the product of velocity and mass, meaning it is the product of velocity, density, and volume. Although the term "quantity of motion" suggests a scalar, we know that velocity is a vector, (i.e., it has a magnitude and a direction), so it's clear that momentum as Newton defined it is also is a vector. After going on to define various kinds of forces and the attributes of those forces, Newton then, took the law of inertia and relativity as his First Law of Motion, just as Descartes and Huygens had done. Following this we have the "force law", i.e., Newton's Second Law of Motion:

The change of motion is proportional to the motive force impressed; and is made in the direction of the right line in which the force is impressed.

Notice that this statement doesn't agree precisely with either of the two forms in which the Second Law is commonly given today, namely, as F = dp/dt or F = ma. The former is perhaps closer to Newton's actual statement, since he expressed the law in terms of momentum rather than acceleration, but he didn't refer to the rate of change of momentum. No time parameter appears in the statement at all. This is symptomatic of a lack of clarity (as in Aristotle's writings) over the distinction between "impulse force" and "continuous force". Recall that our speculative interpretation of Aristotle's downward "weight" was based on the idea that he actually had in mind something like the impulse force that would be exerted by the object if it were abruptly brought to a halt. Newton's Second Law, as expressed in the Principia, seems to refer to such an impulse, and this is how Newton used it in the first few Propositions, but he soon began to invoke the Second Law with respect to continuous forces of finite magnitude applied over a finite length of time - more in keeping with a continuous force of gravity, for example. This shows that even in the final version of the axioms and definitions laid down by Newton, he did not completely succeed in clearly delineating the concept of force that he had in mind. Of course, in each of his applications of the Second

Law, Newton made the necessary dimensional adjustments to appropriately account for the temporal aspect that was missing from the statement of the Law itself, but this was done ad hoc, with no clear explanation. (His ability to reliably incorporate these factors in each context testifies to his solid grasp of the new dynamics, despite the imperfections of his formal articulation of it.) Subsequent physicists clarified the quantitative meaning of Newton's second law, explicitly recognizing the significance of time, by expressing the law either in the form $F = d(mv)/dt$ or else in what they thought was the equivalent form $F = m(dv/dt)$. Of course, in the context of special relativity these two are not equivalent, and only the former leads to a coherent formulation of mechanics. (It's also worth noting that, in the context of special relativity, the concept of force is largely an anachronism, and it is introduced mainly for the purpose of relating relativistic descriptions to their classical counterparts.)

The third Law of Motion in the Principia is regarded by many people as one of Newton's greatest and most original contributions to physics. This law states that

> To every action there is always opposed an equal reaction: or, the mutual actions of two bodies upon each other are always equal, and directed to contrary parts.

Unfortunately the word "action" is not found among the previously defined terms, but in the subsequent text Newton clarifies the intended meaning. He says "If a body impinge upon another, and by its force change the motion of the other, that body also... will undergo an equal change in its own motion towards the contrary part." In other words, the net change in the "quantity of motion" (i.e., the sum of the momentum vectors) is zero, so momentum is conserved. More subtly, Newton observes that "If a horse draws a stone tied to a rope, the horse will be equally drawn back towards the stone". This is true even if neither the horse nor the stone are moving (which of course implies that they are each subject to other forces as well, tending to hold them in place). The illustrates how the concept of force enables us to conceptually decompose a null net force into non-null components, each representing the contributions of different physical interactions.

In retrospect we can see that Newton's three "laws of motion" actually represent the definition of an inertial coordinate system. For example, the first law imposes the requirement that the spatial coordinates of any material object free of external forces are linear functions of the time coordinate, which is to say, free objects move with a uniform speed in a straight line with respect to an inertial coordinate system. Rather than seeing this as a law governing the motions of free objects with respect to a given system of coordinates, it is more correct to regard it as defining a class of coordinates systems in terms of which a recognizable class of motions have particularly simple descriptions. It is then an empirical question as to whether the phenomena of nature possess the attributes necessary for such coordinate systems to exist.

The significance of "force" was already obscure in Newton's three laws of mechanics, but it became even more obscure when he proposed the law of universal gravitation, according to which every particle of matter exerts a force of attraction on every other particle of matter, with a strength proportional to its mass and inversely proportional to the square of the distance. The rival Cartesians expected all forces to be the result of local contact between bodies, as when two objects press directly against each other, but Newton's conception of instantaneous gravity between distant objects seems to defy representation in those terms. In an effort to reconcile universal gravitation with semi-Cartesian ideas of force, Newton's young friend Nicolas Fatio hypothesized an omni-directional flux of small "ultra-mundane" particles, and argued that the mutual shadowing effect could explain why massive bodies are forced together. The same idea was later taken up by Lesage, but many inconsistencies were pointed out, making it clear that no such theory could accurately account for the phenomena. The simple notion of force at a distance was so successful that it became the model for all mutual forces between objects, and the early theories of electricity and magnetism were expressed in those terms. However, reservations about the intelligibility of instantaneous action at a distance remained. Eventually Faraday and Maxwell introduced the concept of disembodied "lines of force", which later came to be regarded as fields of force, almost as if force was an entity in its own right, capable of flowing from place to place. In this way the Maxwellians (perhaps inadvertently) restored the Cartesian ideas that all space must be occupied and that all forces must be due to direct local contact. They accomplished this by positing a new class of entity, namely the field. Admittedly our knowledge of the electromagnetic field is only inferred from the behaviour of matter, but it was argued that explanations in terms of fields are more intelligible than explanations in terms of instantaneous forces at a distance, mainly because fields were considered necessary for strict conservation of energy and momentum once it was recognized that electromagnetic effects propagate at a finite speed.

However, the explanation of phenomena in terms of fields, characterized by partial differential equations, was incomplete, because it was not possible to represent stable configurations of matter in these terms. Maxwell's field equations are linear, so there was no hope of them possessing solutions corresponding to discrete electrical charges or particles of matter. Hence it was still necessary to retain the laws of mechanics of discrete entities, characterized by total differential equations. The conceptual dichotomy between Newton's physics of particles and Maxwell's physics of fields is clearly shown by the contrast between total and partial differential equations, and this contrast was seen (by some people at least) as evidence of a fundamental flaw. In a 1936 retrospective essay Einstein wrote

> This is the basis on which H. A. Lorentz obtained his synthesis of Newton's mechanics and Maxwell's field theory. The weakness of this theory lies in the fact that

it tried to determine the phenomena by a combination of partial differential equations (Maxwell's field equations for empty space) and total differential equations (equations of motions of points), which procedure was obviously unnatural.

The difference between total and partial differential equations is actually more profound than it may appear at first glance, because it entails different assumptions about the existence of free-will and acts of volition. If we consider a point-like particle whose spatial position x(t) is strictly a function of time, and we likewise consider the forces F(t) to which this particle is subjected as strictly a function of time, then the behaviour of this particle can be expressed in the form of total differential equations, because there is just a single independent variable, namely the time coordinate. Every physically meaningful variable exists as one of a countable number of explicit functions of time, and each of the values is realized at it's respective time. Thus the total derivatives are evaluated over actualized values of the variables. In contrast, the partial derivatives over immaterial fields are inherently hypothetical, because they represent the variations in some variable of a particle not as a function of time along the particle's actual path, but transversely to the particle's path. For example, rather than asking how the force experienced by a particle changes over time, we ask how the force would change if at this instant of time the particle was in a slightly different position. Such hypotheticals have meaning only assuming an element of contingency in events, i.e., only if we assume the paths of material objects could be different than they are.

Of course, if we were to postulate a substantial continuous field, we could have non-hypothetical partial derivatives, which would simply express the facts implicit in the total derivatives for each substantial part of the field. However, the intelligibility of a truly continuous extended substance is questionable, and we know of no examples of such a thing in nature. Given that the elementary force fields envisaged by the Maxwellians were eventually concede to be immaterial, and their properties could only be inferred from the state variables of material entities, it's clear that the partial derivatives over the field variables are not only hypothetical, but entail the assumption of freedom of action. In the absence of freedom, any hypothetical transverse variations in a field (i.e., transverse to the actual paths of material entities) would be meaningless. Only actual variations in the state variables of material entities would have meaning. Thus the contrast between total and partial differential equations reflects two fundamentally different conceptual frameworks, the former based on determinism and the latter based on the possibility of free acts. This is closely analogous to Aristotle's dichotomy between natural and violent motions.

As noted above, Einstein regarded this dualism as unnatural, and his intuition led him to expect that the field concept, governed by partial differential equations, would ultimately prove to be sufficient for a complete description of phenomena. In the same essay mentioned above he wrote

> What appears certain to me, however, is that, in the foundations of any consistent field theory, there should not be, in addition to the concept of the field, any concept concerning particles. The whole theory must be based solely on partial differential equations and their singularity-free solutions.

It may seem ironic that he took this view, considering that Einstein was such a staunch defender of strict causality and determinism, but by this time he was wholly committed to the concept of a continuous field as the ultimate ontological entity, more fundamental even than matter, and possessing a kind of relativistic substantiality, subject to deterministic laws. In a sense, he seems to have come to believe that the field was not a hypothetical entity inferred from the observed behaviour of material bodies, but rather that material bodies were hypothetical entities inferred from the observed behaviour of fields. An important first step in this programme was to eliminate the concept of forces acting between bodies, and to replace this with a field-theoretic model. He (arguably) accomplished this for gravitation with the general theory of relativity, which completely dispenses with the concept of a "force of gravity", and instead interprets objects under the influence of gravity as simply proceeding, unforced, along the most natural (geodesic) paths. Thus the concept of force, and particularly gravitational force, which was so central to Newton's synthesis, was simply discarded as having no absolute significance.

However, the concept of force is still very important in physics, partly because we continue to employ the classical formulation of mechanics in the limit of low speeds and weak gravity, but more importantly because it has not proven possible (despite the best efforts of Einstein and others) to do for the other forces of nature what general relativity did for gravity, i.e., to express the apparently forced (violent) motions as natural paths through a modified geometry of space and time.

INERTIAL AND GRAVITATIONAL SEPARATIONS

The special theory of relativity is formulated as a local theory, so its natural focus is on the worldlines of individual particles. In addition, special relativity presupposes a preferred class of worldlines, those representing inertial motion. The idea of a worldline is inherently "absolute" in the sense that it is nominally defined with reference only to a system of space and time coordinates, not to any other objects. This is in contrast to a truly relational theory, which would take the "dual" approach, and regard the separations between particles as the most natural objects of study. In fact, we could go to the relationist extreme of regarding separations as the primary ontological entities, and considering particles to be merely abstract concepts that we use to psychologically organize and coordinate the separations. The relationist view arguably has the advantage of not presupposing a fixed background or even a definite dimensionality of space, since each "separation" could be considered to represent an independent degree of freedom. Of course, this freedom doesn't seem to exist in the real world, since we cannot arrange five particles all mutually equidistant from each other. Indeed it appears that the n(n-1)/2 separations between n particles can be fully encoded as just 3n real numbers, and moreover that those real number vary continuously as the individual particles "move". This is the justification for the idea of particles moving in a coherent three-dimensional space.

Nevertheless, it's interesting to examine the spatial separations that exist between material particles (as opposed to the space and time coordinates of individual particles), to see if their behaviour can be characterized in a simple way. From this point of view, the idea of "motion" is secondary; we simply regard separations as abstract entities having certain properties that may vary with time. In this context, rather than discussing inertial motion of an individual particle, we consider the spatial separation (as a function of time) between two inertial particles. However, since we don't presuppose a background of absolute inertial motion, we will refer to the particles as being "co-inertial", meaning simply that the spatial separation between them behaves like the separation between two particles in absolute inertial motion, regardless of whether the two particles are actually in absolute inertial motion.

Is it possible to characterize in a simple way the spatial separations that exist between co-inertial particles? Consider, for example, the spatial separation s(t) as a function of time between a stationary particle and a particle moving uniformly in a straight line through space, as depicted in the figure below for the condition when the direction of motion of the moving particle B is perpendicular to the displacement from the stationary particle A.

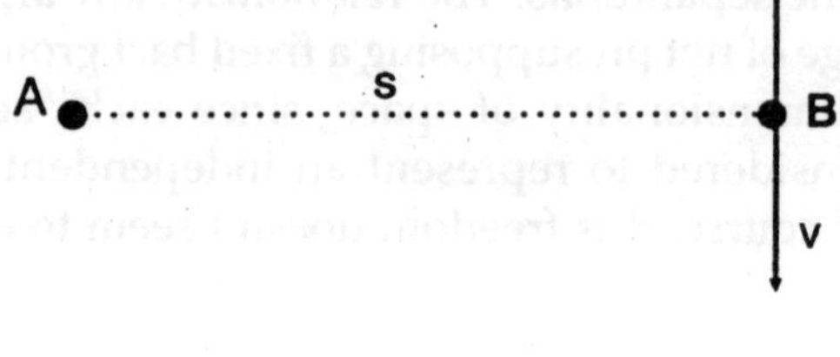

Obviously the separation between objects A and B in this configuration is stationary at this instant, i.e., we have ds/dt = 0, and yet we know from experience that this physical situation is distinct from one in which the two objects are actually stationary with respect to each other's inertial rest frames. For example, the Moon and Earth are separated by roughly a constant distance, and yet we understand that the Moon is in constant motion perpendicular to its separation from the Earth. It is this transverse motion that counter-acts the effect of gravity and keeps the Moon in its orbit. This is another reason that we ordinarily find it necessary to describe motion not in purely relational terms, but in terms of absolutely non-rotating systems of inertial coordinates. Of course, as Mach observed, the apparent existence of "absolute rotation" doesn't necessarily refute relationism as a viable basis for coordinating events. It could also mean that we must take more relations into account. (For example, the Moon's motion is always tangential to the Earth, but it is not always tangential to other bodies, so it's orbital motion does show up in the totality of binary separations.) Whether or not a workable physics could be developed on a purely relational basis is unclear, but it's still interesting to examine the class of co-inertial separations as functions of time. It turns out that co-inertial separations are characterized by a condition that is nearly identical to the condition for linear gravitational free-fall, as well as for certain other natural kinds of motion.

The three orthogonal components Δx, Δy, and Δz of the separation between two particles in unaccelerated motion relative to a common reference frame must be linear functions of time, i.e.,

$$\Delta x = a_1 + b_1 t \qquad \Delta y = a_2 + b_2 t \qquad \Delta z = a_3 + b_3 t$$

where the coefficients a_i and b_i are constants. Therefore the magnitude of any "co-inertial separation" is of the form

$$S(t) = \sqrt{At^2 + Bt + C}$$

where

$$A = b_1^2 + b_2^2 + b_3^2 \qquad B = 2(a_1b_1 + a_2b_2 + a_3b_3)$$
$$C = a_1^2 + a_2^2 + a_3^2$$

Letting the subscript n denote nth derivative with respect to time, the first two derivatives of s(t) are

$$s_1 = \frac{2At + B}{2s_0} \qquad s_2 = \frac{4Ac - B^2}{4s_0^3}$$

The right hand equation shows that $s_2 s_0^3 = k$, and we can differentiate this again and divide the result by s_0^2 to show that the separation s(t) between any two particles in relatively unaccelerated (i.e., co-inertial) motion in Galilean spacetime must satisfy the equation

$$s_0s_3 + 3s_1s_2 = 0 \qquad ...(1)$$

Now we consider the separation that characterizes an isolated non-rotating two-body system in gravitational free-fall. Assume the two bodies are identical particles, each of mass m. According to Newtonian theory the inertial and gravitational constraints are coupled together by the auxiliary quantity called "force" by the following equations

$$F = m\frac{s_2}{2} \qquad F = -G\frac{m^2}{s_0^2}$$

where G is a universal constant. (Note that each particle's "absolute" acceleration is half of the second derivative of their mutual separation with respect to time.) Equating these two forces gives $s^2s_0^2 = -2Gm$. Differentiating this again and dividing by s0, we can characterize non-rotating gravitational free-fall by the purely kinematic equation

$$s_0s_3 + 2s_1s_2 = 0 \qquad ...(2)$$

The formal similarity between equations (1) and (2) is remarkable, considering that the former describes strictly inertial separations and the latter describes gravitational separations. We can show how the two are related by considering general free motion in a gravitational field. The Newtonian equations of motion are

$$r = -\frac{m}{r^2} + r\omega^2 \qquad \ddot{\omega} = -\frac{2}{r}\dot{r}\omega$$

where r is the magnitude of the distance from the centre of the field and w is the angular velocity of the particle. If we solve the left hand equation for w and differentiate to give dw/dt, we can substitute these expressions into the right hand equation and re-arrange the terms to give

$$r\dddot{r} + 3\dot{r}\ddot{r} = -\frac{m}{r^2}\dot{r}$$

which applies (in the Newtonian limit) to arbitrary free paths of test particles in a gravitational field. Obviously if m = 0 this reduces to equation (1), representing free inertial separations, whereas for purely radial motion we have $d^2r/dt^2 = -m/r^2$, and so this reduces to equation (2), representing radial gravitational separation.

Other classes of physical separations also satisfy a differential equation similar to (1) and (2), mentioned earlier. For example, consider a particle of mass m attached to a rod in such a way that it can slide freely along the rod. If we rotate the rod about some point P then the particle in general will tend to slide outward along the rod away from the centre of rotation in accord with the basic equation of motion

$$s_2 - \omega^2 s_0 = 0$$

where s is the distance from the centre of rotation to the sliding particle, and w is the angular velocity of the rod. Differentiating and multiplying through by s0 gives

$$s_0 s_3 - \omega^2 s_0 s_1 = 0$$

Then since $s_2 = \omega^2 s_0$, we see that s(t) satisfies the equation

$$s_0 s_3 + (-1) s_0 s_1 = 0 \qquad \text{...(3)}$$

So, we have found that arbitrary co-inertial separations, non-rotating gravitational separations, and rotating radial separations are all characterized by a differential equation of the form

$$s_0 s_3 + N s_0 s_1 = 0 \qquad \text{...(4)}$$

for some constant N. (Among the other solutions of this equation (with N = -1) are the elementary transcendental functions e^t, sin(t), and cos(t).) Solving for N, to isolate the arbitrary constant, we have

$$\frac{s_0 s_3}{s_1 s_2} = -N$$

Differentiating this gives the basic equation

$$s_0 s_1 s_2 s_3 - s_0 s_1 s_3^{\,2} - s_0 s_2^{\,2} s_3 + s_1^{\,2} s_2 s_3 = 0$$

If none of s_0, s_1, s_2, and s_3 is zero, we can divide each term by all of these to give the interesting form

$$\frac{s_4}{s_3} - \frac{s_3}{s_2} - \frac{s_2}{s_1} + \frac{s_1}{s_0} = 0$$

This could be seen as a (admittedly very simplistic) "unification" of a variety of physically meaningful spatial separation functions under a single equation. The "symmetry breaking" that leads to the different behaviour in different physical situations arises from the choice of N, which appears as a constant of integration.

Incidentally, even though the above has been based on the Galilean spatial separations between objects as a function of Galilean time, the same conditions can be shown to apply to the absolute spacetime intervals between inertial particles as a function of their proper times. Relative to any point on the worldline of one particle, the four components Δt, Δx, Δy, and Δz of the absolute interval to any other inertially moving particle are all linear functions of the proper time τ along the latter particle's worldline. Therefore, the components can be written in the form

$$\Delta t = a_0 + b_0 \Delta\tau \qquad \Delta x = a_1 + b_1 \Delta\tau$$

$$\Delta y = a_2 + b_2 \Delta\tau \qquad \Delta z = a_3 + b_3 \Delta\tau$$

where the coefficients a_i and b_i are constants. It follows that the absolute magnitude of any "co-inertial separation" is of the form

$$s(\tau) = \sqrt{\Delta t^2 - \Delta x^2 - \Delta y^2 - \Delta z^2} = \sqrt{A\tau^2 + B\tau + C}$$

where

$$A = b_0^{\,2} - b_1^{\,2} - b_2^{\,2} - b_3^{\,2}$$

$$B = 2\left(a_0 b_0 - a_1 b_1 - a_2 b_2 - a_3 b_3\right)$$

$$C = a_0^{\,2} - a_1^{\,2} - a_2^{\,2} - a_3^{\,2}$$

Thus we have the same formal dependence as before, except now the parameter s represents the absolute spacetime separation. This shows that the absolute separation between any fixed point on one inertial worldline and a point advancing along any other inertial worldline satisfies equation (1), where subscripts denote derivatives with respect to proper time of the advancing point. Naturally the reciprocal relation also holds, as well as the absolute separation between two points, each advancing along arbitrary inertial worldlines, correlated according to their respective proper times.

WHAT IS RELATIVITY?

Classical relativity involves a simple transformation between a moving object and an observer in another inertial frame of reference. If you are walking in a moving train, and someone stationary on the ground is watching, your speed relative to the observer will be the sum of your speed relative to the train and the train's speed relative to the observer. You're in one inertial frame of reference, the train itself (and anyone sitting still on it) are in another, and the observer is in still another.

The problem with this is that light was believed, in the majority of the 1800s, to propagate as a wave through a universal substance known as the ether, which would have counted as a separate frame of reference (similar to the train in the above example). The famed Michelson-Morley experiment, however, had failed to detect Earth's motion relative to the ether and no one could explain why. Something was wrong with the classical interpretation of relativity as it applied to light ... and so the field was ripe for a new interpretation when Einstein came along. Although he is regarded as one of the most brilliant mathematical physicists of the century, Einstein thought of himself as much as a philosopher as a scientist. Certainly his theories relating matter, energy, space, time and gravity have guided much of the work in theoretical physics since 1905. His famous "thought experiments," based on intuition and imagination rather than laboratory work, propelled us beyond the mechanistic, unchanging "clockwork universe" of Newton and the other classical physicists into a relativistic universe. Here clocks run slower or faster depending on the speed of travel or location in the universe, and "true" distances are stretched or shrunk by gravity.

Einstein's legacy is a universe in which space and time are woven into a single fabric – spacetime. It is matter that causes spacetime to curve and whose motion and properties are, in turn, altered by that curvature.

Einstein proposed two theories of relativity- Special Relativity in 1905, and General Relativity in 1915. Although general relativity is the theory that we will most closely consider in the following topics, Special Relativity is important to understand as it defines much of what we now consider basics of physics and sets a base for General Relativity. Special Relativity is a theory on the nature of matter moving through space and time, and it combines these two properties into a continuum called Space-Time. The heart of the theory is that different observers moving at different rates of speed will find that the laws of physics and the speed of light remain the same, even though they will have different perceptions of time and distance. Because of this, simultaneous events can appear to happen at different times, or events happening hours apart to one person could happen at the same time to another. In other words, your reality is determined by your point of view.

Other important conclusions came from special relativity, including the fact that a moving object is shortened in the direction of movement, time runs more slowly for a moving object, no body can travel at or above the speed of light, and that a body has more mass when it is moving than when it is at rest. These effects are usually too small to be observed with our normal velocities.

Einstein's famous equation, $E=mc^2$, also stems from special relativity. The greater mass of a moving body suggests a connection between kinetic energy and mass. From there, Einstein guessed that all energy and mass would be related. $E=mc^2$ shows this relationship. General Relativity is an extension of special relativity, but it is mainly a theory of gravity. It defines gravity as a curvature of the four-dimensional space-time suggested by special relativity.

The best way to picture this would be to imagine space-time as a rubber sheet. If you were to drop a lead ball onto the sheet, it would curve downwards. This indentation would be the ball's gravitational field. A heavier ball would create a larger 'field'. If you were to drop a marble or much lighter lead ball (because a very dense marble would create a 'field' of it's own!) Near one of the fields, it would roll in, much as a small celestial object would be pulled in by a larger one.

There are, of course, many other conclusions based on general relativity. General relativity gives a description of how material bodies will move in the presence of gravitational fields, and the mathematical equation gives the precise value for curvature caused by bodies of different masses. It predicts that extremely massive stars can collapse on themselves to form Black Holes. General relativity shows that gravity can bend light, such that some of the stars we see in the night sky are not readily visible to us, but the sun curves their light so we can see them. And finally, general relativity shows that a straight line is not always the shortest way between two points. A geodesic is what the line signifying this distance is known as. In Euclidean, or plane, geometry, the geodesic is a straight line. But when relativity comes into play, geodesics can become curved. (Think of the circle routes used by planes on long or overseas flights.)

Such are the basics of relativity. By the way, have you ever wondered why it's called relativity? It's because it all depends on the relative motion of the observers.

Our focus is on General Relativity, but to understand that, you must have basic knowledge of Special Relativity.

Space-Time: A look at the continuum and curvature of space and time.

Black Holes

Wormholes and Time Travel

PARALLEL UNIVERSES

Relativity views space-time as a continuum; that is, time becomes another dimension of space, making space four-dimensional. This is really common sense-we know that it takes time to move through the physical dimensions of space. (length, width, and heigth) Time is a property of space that is distinct from yet closely related to the other three. (Althought we know we can move through space in any direction, we can only move forward in time, at least until wormholes are discovered.)

Curvature of space-time can be more difficult to understand. Although the rubber sheet model gives a picture of how it happens, one cannot easily picture the so-called nothingness of space being curved or a non-physical property like time being warped. But relativity (a theory which has yet to be disproved) predicts this curvature and uses gravity as proof of it. In fact, relativity predicts singularities, points in space-time where the curvature reaches infinity. Singularities are the centres of black holes, points of gravity so strong that nothing, not even light, can escape.

According to relativity, the curvature of space-time is gravity, and this raises an intersting paradox: the amount of curvature is governed by the distribution of matter and energy in the univers, but this distribution is determined by the curvature of space-time. Thinking back to the rubber sheet model, this becomes an obvious truth. A large object, such as a shot put, would cause a large indentation that would cause a marble to roll in. Or, if an area of rubber sheet space contained only small objects, several marbles might roll together to form an area of 'strong gravity', which would in turn attract more objects.

Besides creating the all-important force of gravity, space-time curvature also makes a major dent (no pun intended) in one of our most common beliefs from geometry and family vacations: in curved space-time, a straight line is NOT always the shortest distance between two points. But that's a different topic.

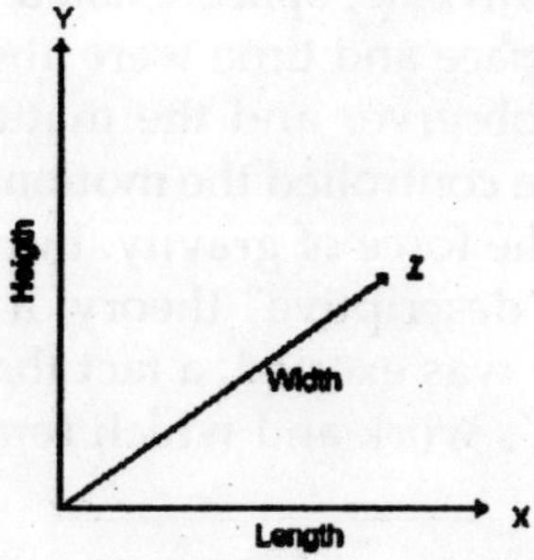

In order to move through the physical dimensions of space, one must also move through time. In this graph, time surrounds the axes, and although the lines could be extended so that we could move in any physical direction, we do not extend them because we do not know how to move backwards in time.

FOUNDATIONS OF RELATIVITY FOR RELATIVE BEGINNERS

Einstein's Special Theory of Relativity challenged long-held notions about space and time that had been established over two centuries earlier by Isaac Newton.

One of the foremost pioneers of modern science, Isaac Newton developed his three laws of motion and a theory of gravity, not to mention the calculus needed to develop and express these theories in math! He set his concepts in a framework of space and time which he (like everyone else at that time) assumed to be absolute.

For two centuries that omission was overshadowed by his triumphs in celestial mechanics and optics.

What, in essence, were the key concepts in Newton's theory of gravitation? Masses experience an attractive force between them, a force which acts at a distance, resulting in their acceleration toward each other. The strength of that force depends on the size of the masses and is inversely proportional to square of the distance between them.

In Newton's universe, space existed independent of the matter in it. Both space and time were absolute, regardless of the motion of the observer and the matter contained within space. No substance controlled the motions of the moon, Earth and planets; only the force of gravity. But Newton's theory of gravitation was a "descriptive" theory; it didn't explain how the force of gravity was exerted, a fact that has an interesting parallel in Einstein's work and which remains a hot topic for future research.

Newton's laws satisfactorily explained most phenomena studied for the next two hundred years. Toward the end of the 19th century, however, as measuring devices grew more and more precise, the list of puzzling inconsistencies was growing.

Between the 18th and 19th centuries, several problems arose in astronomy and physics that would later challenge Newton's universe.

MERCURY'S TIME-KEEPING

The first problem concerned the planet Mercury. In the late 1800's, astronomers noted that there were slight differences between Mercury's observed orbit and that predicted by Newton's theory of gravitation. Mercury's elliptical path around the Sun shifts slightly forward with each orbit such that its closest point ("perihelion") advances by 5 arc minutes per century. Newton's theory could explain all but 42 arc seconds. No one could account for this tiny but significant discrepancy until Einstein.

The second area of theoretical disquiet concerned the brilliant new unified theory of electromagnetism, developed in the 1870's by the Scottish physicist James Clerk Maxwell. His theory, which predicted the existence of electromagnetic waves moving at the speed of light, was highly successful in all experimental tests. But it seemed to require that such waves move through a fixed reference frame which, in contrast to Newton's empty space, was believed at that time to consist of an invisible substance known as the aether. In 1887, Albert A. Michelson and Edward W. Morley tried to measure the speed of the ether. The concept of the ether was made in analogy with other types of media in which different types of waves are able to propagate; sound waves can, for example, propagate in air or other materials. The result of the Michelson-Morley experiment was that the speed of the Earth through the ether (or the speed of the ether wind) was zero. Therefore, this experiment also showed that there is no need for any ether at all, and it appeared that the speed of light in vacuum was independent of the speed of the observer! Michelson and Morley repeated their experiment many times up until 1929, but always with the same results and conclusions. Michelson won the Nobel Prize in Physics in 1907.

When describing systems at rest in absolute space, Maxwell's laws were simple and elegant, but when they were applied to moving systems, the mathematics grew much more complex and difficult to interpret. Scientists, particularly Einstein, began to explore Maxwell's laws further. Its equations turned out to be essentially correct, but Einstein attempted to explain why in his Special Theory of Relativity.

A third problem in the Newtonian universe concerned the speed of light. To an observer at rest in Newton's absolute space, light travels (through the aether) at a constant speed. But logic dictates that when the observer is moving through space, the observed speed of light should vary, depending on whether the observer is moving toward the source of light or away from it.

By the end of the 19th century, more sensitive instrumentation allowed this notion to be put to the test. In 1881, using a very sensitive apparatus, Albert Michelson found no difference at all between the speeds of light measured for opposite directions of travel. Working with chemist Edward Morley in 1887, he repeated the experiment using yet more accurate instruments; they obtained the same results. The truth of observations were undeniable. Scientists began to explain the results by theorizing that distance through which light travelled might appear to stretch or shrink, depending on the direction of travel.

The notion of the aether as a fixed reference frame through which light must propagate was beginning to unravel. No experiments had demonstrated its existence. Einstein's Special Theory of Relativity was to dispense with the aether altogether.

As a young man in college, Einstein seized upon these theoretical difficulties as interesting puzzles to probe. His interest in them did not wane when he failed to get a teaching post and eventually secured a position with the Swiss patent office. Einstein continued to study the new physics on his own time and in relative isolation, though he enthusiastically discussed philosophical and scientific topics with a circle of college friends. He was working on several ideas at once, including his Ph.D. thesis research and a separate paper on special relativity.

GALILEAN RELATIVITY

Relativity can be described using space-time diagrams. Contrary to popular opinion, Einstein did not invent relativity. Galileo preceded him. Aristotle had proposed that moving objects (on the Earth) had a natural tendency to slow down and stop. This is shown in the space-time diagram below.

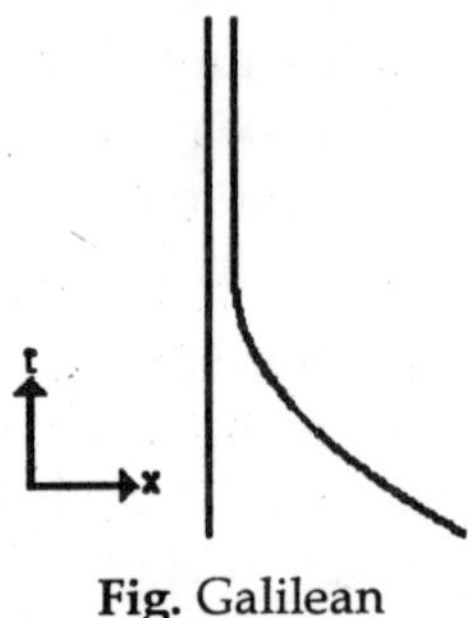

Fig. Galilean Relativity Diagram

Note the curved worldline above. This shows a variable velocity, or an acceleration. Galileo objected to Aristotle's hypothesis, and asked what happened to an object moving on a moving ship:

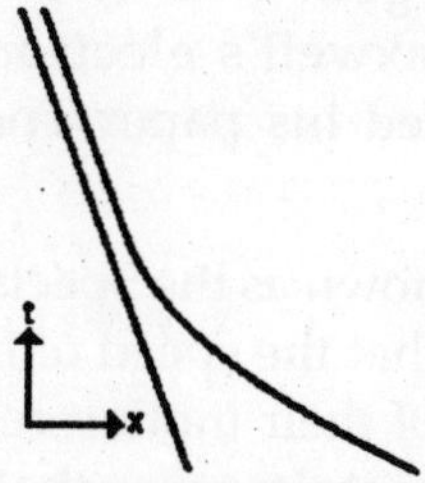

Fig. Galilean Relativity Diagram

Now it is still moving in its final state. Galileo proposed that it is only relative velocities that matter. Thus a space-time diagram can be transformed by painting it on the side of a deck of cards, and then skewing the deck to one side – but keeping the edges along a straight line:

Straight worldlines (unaccelerated particles) remain straight in this process. Thus Newton's First Law is preserved, and non-accelerated worldlines are special. This Galilean transformation does not affect the time. Thus two observers moving with respect to each other can still agree on the time, and thus the distance between two objects, which is the difference in their positions measured at equal times, can be defined. This allowed Newton to describe an inverse square law for gravity.

But Galilean transformations do not preserve velocity. Thus the statement "The speed limit is 70 mph" does not make sense – but don't try this in court. According to relativity, this must be re-expressed as "The magnitude of the relative velocity between your car and the pavement must be less than 70 mph".

WHAT'S SO SPECIAL ABOUT RELATIVITY?

Einstein's first theory of relativity, which he published in 1905, broke away from the Newtonian reliance on space and time as immutable frames of reference. This theory was immediately recognized by the scientific community as having profound implications for physics and cosmology.

Einstein's main goal was to address the apparent inconsistencies in Maxwell's electromagnetic theory. No wonder Einstein named his paper The Electrodynamics of Moving Bodies.

Later to become known as the Special Theory of Relativity, its first postulate was that the speed of light is the same for all observers, regardless of their motion relative to the source of the light. The second postulate was that all observers moving at constant speed should observe the same physical laws. Putting these two ideas together, Einstein showed that the only way this can happen is if time intervals and/or lengths change according to the speed of the system relative to the observer's frame of reference. This flies against our everyday experience but has since been demonstrated to hold in a number of very solid experiments. For example, scientists have shown that an atomic clock travelling at high speed in a jet plane ticks more slowly than its stationary counterpart.

Einstein's discovery of the relativity of space and time led to an equally revolutionary insight. Matter and energy are interrelated, even equivalent. The equivalence of matter and energy is summed up in the famous equation:

$E = mc$, where m = mass and c = the speed of light

Einstein's 1905 theory is referred to as the "special" theory because it is limited to bodies moving in the absence of a gravitational field. It took Einstein eleven more years to formulate a set of general laws that took account of gravity. The result: Einstein's second landmark paper on General Relativity.

THE POSTULATES OF SPECIAL RELATIVITY: NUTSHELL

The theory of special relativity can be derived formally from a small number of postulates. The fundamental postulates of special relativity can be expressed in various ways; you may find different versions of them in different books.

The first two postulates below are assertions about the structure of spacetime, while the last two postulates form the heart of special relativity.

The Geometry of Spacetime

Statement: "Space and time form a 4-dimensional continuum".

The postulate that spacetime forms a 4-dimensional continuum is a generalization of the classical Galilean concept that space and time form separate 3 and 1 dimensional continua. The postulate of a 4-dimensional spacetime continuum is retained in general relativity.

Physicists widely believe that this postulate must ultimately break down, that space and time are quantized over extremely small intervals of space and time, the Planck length $(G\ hbar/c^3)^{1/2}$ = approx 10^{-35} metres, and the Planck time $(G\ hbar/c^5)^{1/2}$ = approx 10^{-43} seconds, where G is Newton's gravitational constant, hbar = h/(2 pi) is Planck's constant divided by 2 pi, and c is the speed of light.

The Existence of Globally Inertial Frames

Statement: "There exist global spacetime frames with respect to which unaccelerated objects move in straight lines at constant velocity".

A spacetime frame is a system of coordinates for labelling space and time. Four coordinates are needed, because spacetime is 4-dimensional.

A frame in which unaccelerated objects move in straight lines at constant velocity is called an inertial frame. One can easily think of non-inertial frames: a rotating frame, an accelerating frame, or simply a frame with some bizarre Dahlian labelling of coordinates.

A globally inertial frame is an inertial frame that covers all of space and time.

The postulate that globally inertial frames exist is carried over from classical mechanics (Newton's first law of motion).

Implicit in the assumption of the existence of globally inertial frames is the assumption that the geometry of spacetime is flat, the geometry of Euclid. In general relativity, this postulate is replaced by the weaker postulate that local (not global) inertial frames exist. A locally inertial frame is one which is inertial in a 'small neighbourhood' of a spacetime point. In general relativity, spacetime can be curved.

The Speed of Light is Constant

Statement: "The speed of light c is a universal constant, the same in any inertial frame".

This postulate is the nub of special relativity, and much of the content of these pages is concerned with exploring its paradoxical consequences, starting with the next page, The Paradox of Special Relativity. Amongst other strange consequences, the postulate implies that the time dimension behaves in many ways as if it were an imaginary spatial dimension. Watch this 4D spacetime hypercube rotate (47K GIF movie).

Measuring speed requires being able to measure intervals of both space and time: speed is distance travelled divided by time elapsed. Inertial frames constitute a special class of spacetime coordinate systems; it is with respect to distance and time intervals in these special frames that the speed of light is asserted to be constant.

In general relativity, arbitrarily weird coordinate systems are allowed, and light need move neither in straight lines nor at constant velocity with respect to bizarre coordinates (why should it, if the labelling of space and time is totally arbitrary?). However, general relativity asserts the existence of locally inertial frames, and the speed of light is a universal constant in those frames.

In 1983, the General Conference on Weights and Measures officially defined the speed of light to be c = 299,792,458 metres/second, and the metre, instead of being a primary measure, became a secondary quantity, defined in terms of the second and the speed of light.

The Principle of Special Relativity

Statement: "The laws of physics are the same in any inertial frame, regardless of position or velocity".

Physically, this means that there is no absolute spacetime, no absolute frame of reference with respect to which position and velocity are defined. Only relative positions and velocities between objects are meaningful.

It is to be noted that the Principle of Special Relativity does not imply the constancy of the speed of light, although the postulates are consistent with each other. Moreover the constancy of the speed of light does not imply the Principle of Special Relativity, although for Einstein the former appears to have been the inspiration for the latter.

As discussed in the laws of physics are generally assumed to be invariant under displacements in time and space, as well as rotations about any axis. Einstein went one step further and advocated the following additional symmetry for the laws of physics.

Chapter 8

Special Relativity

In his essays: "The Theory of Relativity and other Essays" 1996, Einstein "answered" questions about relativity as follows:

"The theory of relativity is that physical theory which is based on a consistent physical interpretation of the concepts of motion, space and time.

"The name `theory of relativity' is connected with the fact that motion from the point of view of possible experience always appears as the relative motion of one object with respect to another. Motion is never observable as 'motion with respect to space' or, as it has been expressed, as 'absolute motion'.

"The 'principle of relativity' in it's widest sense is contained in the statement: The totality of physical phenomena is of such a character that it gives no basis for the introduction of the concept of 'absolute motion'; or shorter but less precise: There is no absolute motion.

"The development of the theory of relativity proceeded in two steps, 'special theory of relativity' and 'general theory of relativity'. The latter presumes the validity of the former as a limiting case and is its consistent continuation."

It is worth to emphasize the fact that Einstein stated above - it is not two theories: general relativity completely includes special relativity as a limiting case where the gravitational field is negligible. Such a limiting case can happen in empty space, far away from massive objects and in the absence of acceleration. It can even happen closer to massive objects, inside a small laboratory that is in free-fall.

200 years after Newton the theory of electromagnetism was developed into Maxwell's equations. These equations describe waves with a speed of 1/sqrt(epsilon$_o$*mu$_o$), where epsilon$_o$ is the constant describing the strength of the electrostatic force in a vacuum, and mu$_o$ is the constant describing the strength of the magnetic interaction in a vacuum. This is an absolute velocity — it is not relative to anything. The value of the velocity was very close to the measured speed of light, and when Hertz generated electromagnetic waves (microwaves) in his laboratory and showed that they could be reflected and refracted just like light, it became clear that light was just an example of electromagnetic radiation. Einstein tried to fit the idea of an absolute speed of light into Newtonian mechanics. He found that the transformation from one reference frame to another had to affect the time — the idea of sliding a deck of cards had to be abandoned. This led to the theory of special relativity. In special relativity, the velocity of light is special. Anything moving at the speed of light in one reference frame will move at the speed of light in all unaccelerated reference frames. Other velocities are not preserved, so you can still try to get lucky on speeding tickets.

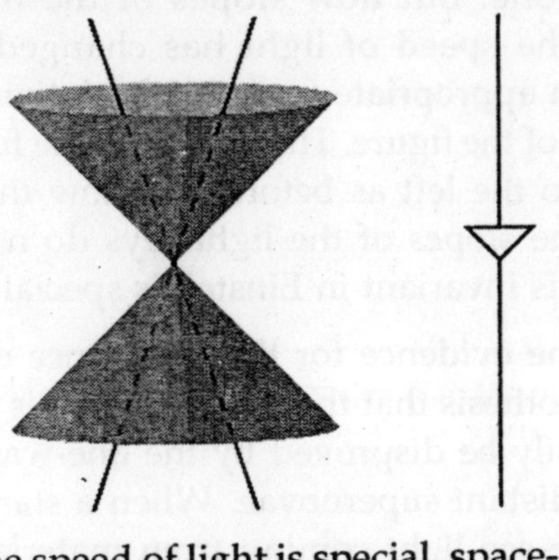

Because the speed of light is special, space-time diagrams are often drawn in units of seconds and light-seconds, or years and light-years, so a unit slope [45 degree angle] corresponds to the speed of light. The set of all light speed world lines going through an event defines the light cones of that event: the past light cone and the future light cone. An example of light cones is shown above. The fancy light picture on the left shows both the past and future light cones of the event where the two worldlines cross, while the schematic version on the right is easy to use in more complicated diagrams.

Thus in the situation shown in 3 space-time diagrams below, the central section shows the worldline of one stationary observer, one observer moving to the right, and two events on the future light cone on the event where the two observers' worldlines cross.

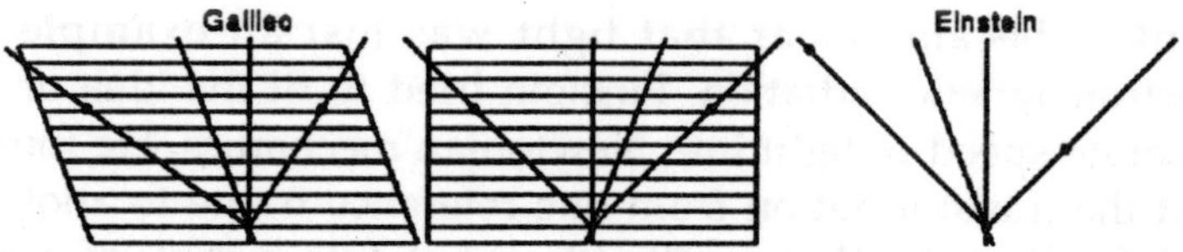

The left-hand section of the figure shows the Galilean transformation into the frame of reference of the moving observer. The events on the future light cone have shifted to the left, but they are still at the same time. Since the coordinates x and t just provide a way of describing space-time, and are not the space-time themselves, the two events are still on the future light cone. But now slopes of the light rays have changed, so the speed of light has changed. The Lorentz transformation appropriate for special relativity is shown on the right hand of the figure. The events on the future light cone have shifted to the left as before, but now their times have changed, so the slopes of the light rays do not change. The speed of light is invariant in Einstein's special relativity.

What is the evidence for the invariance of the speed of light? The hypothesis that the speed of light is c relative to its source can easily be disproved by the one-way transmission of light from distant supernovae. When a star explodes as a supernova, we see light coming from material with a large range of velocities dv, at least 10,000 km/sec. Because of this range of velocities, the spectral lines of a supernova are very broad due to the Doppler shift. After traveling a distance D in time D/c, the arrival time of the light would be spread out by dt = (dv/c)(D/c).

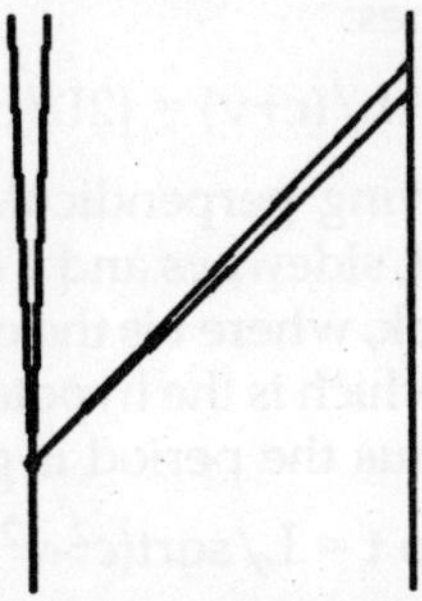

However, this DOES NOT happen. For the Crab supernova, with D/c = 6000 years, dv = 10,000 km/sec would give a range of arrival times of 200 years. But the Crab was only bright for 1 year. For very distant supernovae with D/c = 5 billion years, modern observations with spectrographs show that the redshifted and blueshifted light arrives at the same time: within 10 days. This limit on the spread is 5 billion times smaller than the prediction of the "bullet" model of light.

However, light could travel at speed c relative to a medium — the ether. If it did, then the rate of a "bouncing photon clock" moving with respect to the ether.

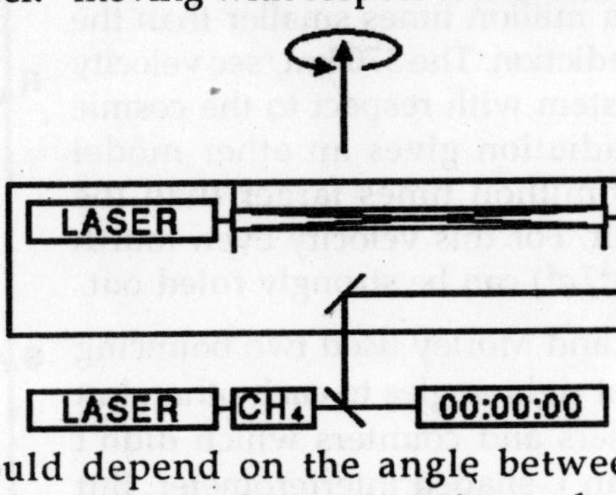

It would depend on the angle between its photon bouncing axis and its velocity. A stationary bouncing photon clock has a period of P = 2L/c. If it is moving parallel to its axis at velocity v, and light moves at speed c with respect to the ether, then the speed relative to the clock when the photon is moving "upstream" is c-v and the one-way time is L/(c-v). When the photon is moving "downstream" the speed relative to the clock is c+v so the one-way time is L/(c+v). The period is the sum of these times:

$$P(par) = L/(c-v) + L/(c+v) = [2L/c]/(1-v^2/c^2).$$

If the clock is moving perpendicular to its axis, the light has to move a distance L sideways and a distance vt "upstream" to keep up with the clock, where t is the one-way time. The total distance traveled is ct, which is the hypotenuse of a right triangle with sides L and vt. Thus the period is given by:

$$(ct)^2 = L^2 + (vt)^2 \text{ so } t = L/sqrt(c^2-v^2)$$

$$P(perp) = 2t = [2L/c]/sqrt(1-v^2/c^2).$$

Thus the ether model predicts that

$$dP/P = [P(par)-P(perp)]/P = 0.5*v^2/c^2.$$

Brillet and Hall (1979, PRL, 42, 549) actually built a bouncing photon clock (a laser stabilized to a Fabry-Perot etalon) on a rotating table and compared its rate to an atomic clock (a laser stabilized to a methane line).

The observed dP/P was $(1.5 +/- 2.5)*10^{-15}$. For the minimum possible velocity of 30 km/sec, due to the orbit of the Earth around the Sun, this is at least a million times smaller than the ether model prediction. The 370 km/sec velocity of the Solar System with respect to the cosmic background radiation gives an ether model prediction 100 million times larger than the Brillet-Hall limit. For this velocity even fourth order effects (v^4/c^4) can be strongly ruled out.

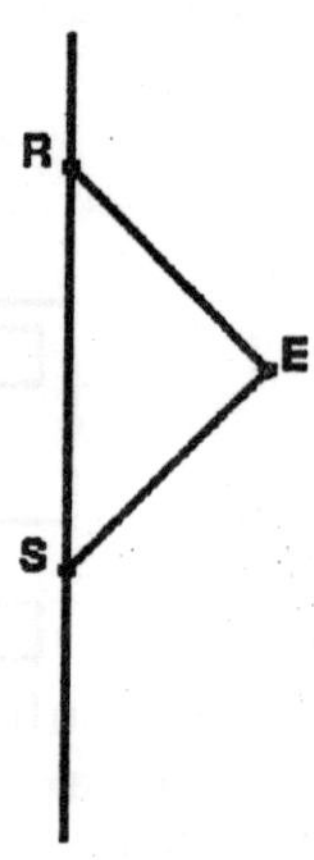

Michelson and Morley used two bouncing photon clocks at right angles to each other, but without the lasers and counters which didn't exist. This left an L-shaped interferometer. But they were able to show that dP/P was essentially zero instead of the ether model prediction.

If we send out a radar pulse at time ts, which is reflected at the event E, and the echo arrives back at time tr, then we know that the light was traveling at c for the entire round-trip journey out to E and back, so the distance of E is D(E) = c*(tr-ts)/2. In our frame of reference we are stationary, and light travels at the same speed coming back from E as it did going out, so the time of the event E is obtained by averaging the send and receive times, t(E) = (tr+ts)/2.

TIME DILATION

Armed with radar, we can determine the time of two events on the worldline of an observer moving with respect to us. We can then compare the time interval we measure to the time interval measured by the moving observer. Consider the two observers A and B below.

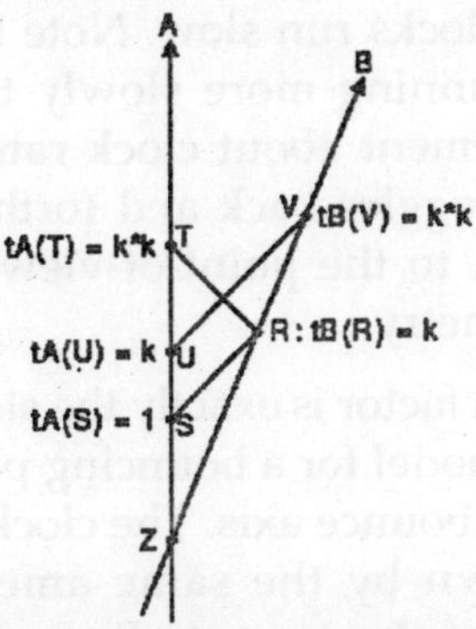

They both set their clocks to zero at the event Z where their worldlines cross. A sends out a radar pulse at the event S at time tA(S) = 1. This is received by observer B at event R with time tB(R) = k. The factor k depends on the relative velocity of A and B, but since the light takes some time to travel between A and B, we know that k will be larger than 1. The notation tA means times determined by A, while tB are times determined by B. If A sends out a pulse at some other time, tA = x, it will be received by B at time tB = x*k by the principle of similar triangles. In particular, if A sends out a pulse at time tA = k it will be received by B at time tB = k*k. Now consider the radar pulse reflected by B at event R. It starts at time tB = k. When will A receive it? Since the speed of A with respect to B is the same as the speed of B with respect to A, this time has to be tA(T) = k*k. We can now compute the distance of event R from A's worldline and time of event R according to A, tA(R). These are DA(R) = c(k*k-1)/2 and tA(R) = (k*k+1)/2. Thus the speed of B according to A is

v = D/t = DA(R)/tA(R) = c(k*k-1)/(k*k+1).

We can solve for k giving

k = sqrt((1+v/c)/(1-v/c))

which is the relativistic Doppler shift formula. But we also find that tA(R) > tB(R), so A says that B's clock is running slow. The amount of this time dilation is

$$(1+k*k)/(2*k) = 1/\mathrm{sqrt}(1-v^2/c^2).$$

Thus moving clocks run slow. Note that B will also find that A's clock is running more slowly than his. There is a symmetric disagreement about clock rates. I have prepared an animation that toggles back and forth from the point-of-view of observer A to the point-of-view of observer B and illustrates the symmetry.

This slow down factor is exactly the slow down calculated above in the ether model for a bouncing photon clock moving perpendicular to its bounce axis. The clock moving parallel to the axis slows down by the same amount under special relativity because of the Lorentz-Fitzgerald contraction of moving objects in the direction of motion.

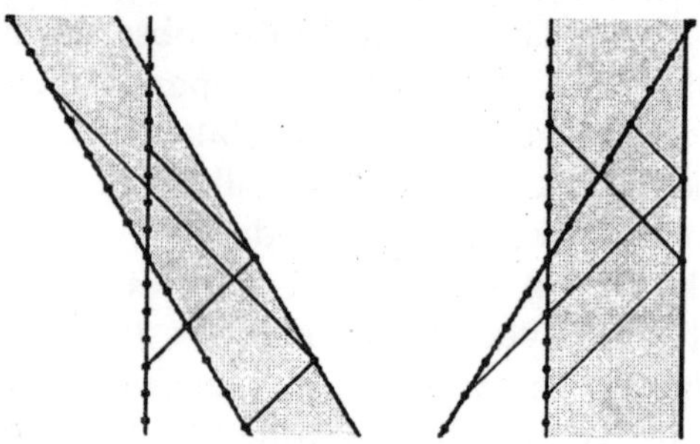

The space-time diagrams above both show a rod moving past an observer. On the left the rod is moving, while on the right the same situation is shown in the rod's frame of reference. The observer moving with respect to the rod makes a radar determination of its length, as does an observer moving along with the rod. The observer on the rod sees a length of 5 light-ticks because it takes 10 ticks for light to make the round trip to the end of the rod and back. The observer moving with respect to the rod at v = 0.6*c measures only 8 ticks for the round trip and thus gives the length of the rod as 4 light-ticks. Thus the length of a moving rod appears to be reduced by a factor of $\mathrm{sqrt}(1-v^2/c^2)$. Thus length contraction changes P(par) for the bouncing photon clock to

$$
\begin{aligned}
P(par) &= [2L*\mathrm{sqrt}(1-v^2/c^2)/c]/(1-v^2/c^2) \\
&= [2L/c]/\mathrm{sqrt}(1-v^2/c^2) = P(perp)
\end{aligned}
$$

so the rate of a bouncing photon clock does not depend on the angle between its velocity and its bouncing axis.

Because the clocks of different observers run at different rates, depending on their velocities, the time for a given observer is a property of that observer and his worldline. This time is called the proper time because it is "owned" by a given particle, not because it is the "correct" time. Proper time is invariant when changing reference frames because it is the property of a particle, not of the reference frame or coordinate system. In general, given any two events A and B with B inside the future light cone of A, there is one unaccelerated worldline connecting A and B, just as there is one straight line connecting two points in space. In the frame of reference of the observer following this unaccelerated worldline, his clock is always stationary, while clocks following any other worldline from A to B will be moving at least some of the time. Because moving clocks run slow, these observers will measure a smaller proper time between events A and B than the unaccelerated observer. Thus the straight worldline between two events has the largest proper time, and all other curved worldlines connecting the two events have smaller proper times. This is exactly analogous to the fact that the straight line between two points has the smallest length of all possible curves between the points. Thus the "twin paradox" is no more paradoxical than the statement that a man who drives straight from LA to Las Vegas will cover fewer miles than a man who drives from LA to Las Vegas via Reno.

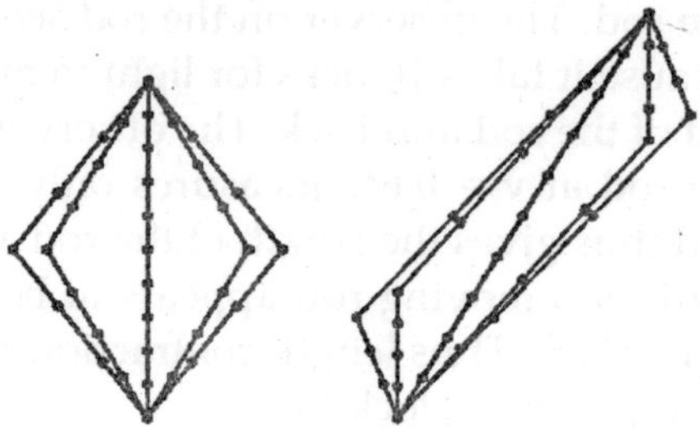

The pair of space-time diagrams above show quintuplets separated at birth. The middle worldline shows the quint who stays home. The space-time diagram on the left is done from the point of view of the middle quint. Each dot on a worldline is a birthday party, so the middle quint is 10 years old when they all rejoin each other, while the other quints are 6 and 8 years old. The space-time diagram on the right shows the same events from the point of view of an observer initially moving with one of the moving quints. When the quints come together their ages are still 6, 8, 10, 8, and 6 years. Thus the straight worldline between two events can be found by maximizing the proper time, just as the straight line between two points can be found by minimizing the length.

EINSTEIN'S RELATIVITY

Newton's laws of motion work very well for anything moving at much less than the speed of light. His law of gravity works very well for any place of weak gravity such as in the solar system. Spacecraft sent to the distant planets in the solar system arrive at their intended destinations (barring mechanical problems) within a few minutes of the expected time even after travelling for billions of kilometers over several years time. The scientists use Newton's laws to guide the spacecraft to its destination. In this chapter you will see what happens when you look at things moving very fast (at a significant fraction of the speed of light) and you will see that space and time can be radically changed in a very strong gravitational field. In fact, you will find that gravity is not really a force!

This revolutionary leap in our understanding of gravity and the nature of space and time was made by Albert Einstein (lived 1879 – 1955). In the first two decades of the 20th century, Einstein laid out a new paradigm of gravity and motion in space and time. In this chapter you will explore his Special Relativity and General Relativity theories that are this new paradigm. The vocabulary terms are in boldface.

Spacetime

Time is not completely separate from and independent of space as you would ordinarily assume. In his Special Relativity theory, Einstein assumed that the fundamental laws of physics do not depend on your location or motion. Two people, one in a stationary laboratory and another in a laboratory aboard a train or rocket moving in a straight line at uniform speed, should get the same results in any experiment they conduct. In fact, if the laboratory in the train or rocket is soundproof and has no windows, there is no experiment a person could conduct that would show he/she is moving.

The laws of physics include the laws of electromagnetism developed by James Maxwell and Maxwell found that electromagnetic waves should travel at a speed given by the combination of two universal constants of nature. Since the laws of physics do not depend on your location or motion, Einstein reasoned that the speed of light will be measured to be the same by any two observers regardless of their velocity relative to each other. For example, if one observer is in a rocket moving toward another person at half the speed of light and both observers measure the speed of a beam of light emitted by the rocket, the person at rest will get the same value the person in the rocket ship measures (about 300,000 kilometers/

second) instead of 1.5 times the speed of light (=rocket speed + speed of beam of light). This assumption has now been shown to be correct in many experiments. To get the same value of the speed (= distance/time) of light, the two observers moving with respect to each other would not only disagree on the distance the light travelled as Newton said, they would also disagree on the time it took.

Einstein found that what you measure for length, time, and mass depends on your motion relative to a chosen frame of reference. Everything is in motion. As you sit in your seat, you are actually in motion around the centre of the Earth because of the rapid rotation of the Earth on its axis. The Earth is in motion around the Sun, the Sun is in orbit around the centre of our Galaxy, the Galaxy is moving toward a large group of galaxies, etc. When you say something has a velocity, you are measuring its change of position relative to some reference point which may itself be in motion. All motion is relative to a chosen frame of reference. That is what the word "relativity" means in Einstein's Relativity theories. The only way observers in motion relative to each other can measure a single light ray to travel the same distance in the same amount of time relative to their own reference frames is if their "metres" are different and their "seconds" are different! Seconds and metres are relative quantities.

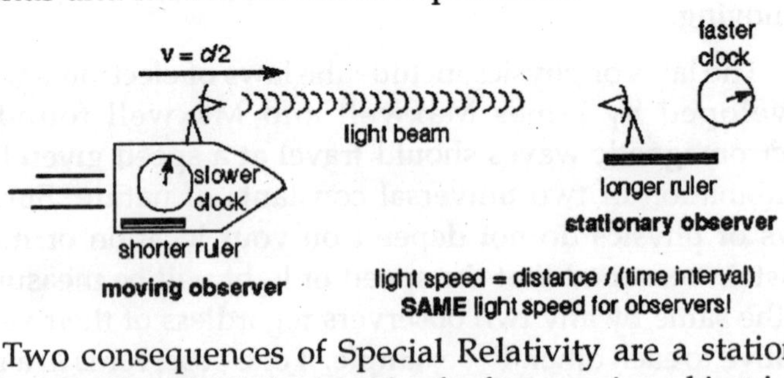

Two consequences of Special Relativity are a stationary observer will find (1) the length of a fast-moving object is less than if the object was at rest, and (2) the passage of time on the fast-moving object is slower than if the object was at rest. However, an observer inside the fast-moving object sees everything inside as their normal length and time passes normally, but all of the lengths in the world outside are shrunk and the outside world's clocks are running slow.

One example of the slowing of time at high speeds that is observed all of the time is what happens when cosmic rays (extremely high-energy particles, mostly protons) strike the Earth's atmosphere. A shower of very fast-moving muon particles are created very high up in the atmosphere. Muons have very short lifetimes – only a couple of millionths of a second. Their short lifetime should allow them to travel at most 600 metres. However they reach the surface after travelling more than 100 kilometers! Because they are moving close to the speed of light, the muons' internal clocks are running much slower than stationary muons. But in their own reference frame, the fast-moving muons's clocks run forward "normally" and the muons live only a couple of millionths of a second.

Time and space are relative to the motion of an observer and they are not independent of each other. Time and space are connected to make four-dimensional spacetime (three dimensions for space and one dimension for time). This is not that strange – we often define distances by the time it takes light to travel between two points. For example, one light year is the distance light will travel in a year. To talk about an event, you will usually tell where (in space) and when (in time) it happened. The event happened in spacetime.

Another consequence of Special Relativity is that nothing can travel faster than the speed of light. Any object with mass moving near the speed of light would experience an increase in its mass. That mass would approach infinity as it reached light speed and would, therefore, require an infinite amount of energy to accelerate it to light speed. The fastest possible speed any form of information or force (including gravity) can operate is at the speed of light. Newton's law of gravity seemed to imply that the force of gravity would instantly change between two objects if one was moved – Newton's gravity had infinite speed (a violation of Special Relativity). The three strange effects of Special Relativity (shrinking lengths, slowing time, increasing mass) are only noticeable at speeds that are greater than about ten per cent of the speed of light. Numerous experiments using very high-speed objects have shown that Special Relativity is correct.

Special Relativity also predicts that matter can be converted into energy and energy in to matter. By applying Newton's second law of motion to the energy of motion for something moving at high speed (its "kinetic energy"), you will find that energy = mass × (speed of light)^2. More concisely, this is Einstein's famous equation, $E = mc^2$. This result also applies to an object at rest in which case, you will refer to its "rest mass" and its "rest energy", the energy equivalent of mass. The amount of rest energy in something as small as your astronomy textbook, for example, is tremendous. If all of the matter in your textbook was converted to energy, it would be enough energy to send a million tons to the Moon!

THE PRINCIPLES OF SPECIAL RELATIVITY

Einstein was lead to his special theory of relativity by his believe that there is no way to detect absolute motion. This dictated that the measured speed of light must be the same in all inertial frames of reference. If this was not so, the laws of physics would have been different in inertial frames that move relative to each other.

Inertial frames are uniformly moving coordinate systems, far away from gravitational effects or any other form of influence, where inertia is isotropic, meaning a given force will cause the same acceleration on identical masses in whatever direction the force is applied.

THE AETHER ABOLISHED

Einstein also argued that if there was a 'rest-frame' for light (the luminiferous aether), we could in principle set up an inertial frame in which light would not propagate in the forward direction at all, e.g., if the frame moves at the speed of light relative to the aether. Einstein was still very young, when he reportedly contemplated if he would still be able to see his own face in a mirror if they were both at rest in such a frame - moving at the speed of light through the aether.

Einstein realized that it is paradoxical to assume the same light ray can actually move with the same speed c (in an absolute Newtonian sense) relative to all inertial frames. This would require that light adapt its "absolute speed" to the frame that measures it. He decided that either time intervals or distance intervals (or both) must change if measured by observers in different inertial frames in relative motion.

THE SPACETIME INTERVAL

In Newtonian dynamics, space intervals and time intervals are the same for all inertial observers, no matter how fast they move. In Einstein's dynamics, it is the spacetime interval that remains the same for all inertial observers, no matter how fast they move. This concept offers the best answer to the question: "what is relativity?"

The square of the spacetime interval is the difference between the square of the time interval and the square of the distance interval, given that all are expressed in the same units. Time is converted to a distance by multiplying it with the speed of light, e.g., one second represents 300 million metres (the speed of light is 300 million metres per second).

Example

Let us calculate a simple example. Suppose you ride on-board a spaceship that travel from Earth to Alpha Centauri at a speed of 80% of the speed of light. The trip takes 3 years on your calendar. But wait, Alpha Centauri is about 4 light-years from Earth, so how could you possible have reached it in 3 years? To answer the question, consider your spacetime interval between the two events - your departure from Earth and your arrival at Alpha Centauri.

The time interval on your calendar is 3 years. The space interval in your reference frame is zero. How can this be? Well, you were present at both events (departure and arrival), so in your frame of reference, there cannot be a space interval between the two events! So what is the spacetime interval? It must be 3 light-years, because you subtract zero-squared from 3-squared, leaving you with 3-squared.

SOLUTION

Now this spacetime interval must remain constant for all observers, thus also for Earth-bound observers. From Earth, we measured the distance to Alpha Centauri accurately as 4 light-years, which must then be the space interval between your departure from here and your arrival at Alpha Centauri. We know that the spacetime interval is 3 years for everyone, so what time interval does it give us? We take 3-squared and add 4-squared, giving 25. Take the square root and we have the time interval for Earth's inertial frame: 5 years.

So, after all, you did not travel at faster than light! In our reference frame, you traveled at our original assumption of 80% of the speed of light (4 light-years in 5 years). If we, erroneously, take our distance and divide it by your time, we would end up with 4 divided by 3, or a speed of 133% of light-speed. If you insist that you want to know the distance between Earth and Alpha Centauri in your inertial reference frame, you can take your measured speed (0.8c) and multiply it by your measured time interval (3 years) and get a distance of 2.4 light-years.

TIME DILATION AND LENGTH CONTRACTION

In trying to answer the question "what is relativity?", we have indirectly 'stumbled upon' the phenomena of time dilation and length contraction due to velocity. The constancy (or invariance) of the spacetime interval is underlying to relativistic time dilation and length contraction. With your relative speed of 0.8c, your time dilation factor is 0.6, meaning you will measure times and radial distances that are only 60% of what we on Earth would have measured.

The reason for belaboring the observation of the time- and space-intervals between events is this: events in empty space give us something 'tangible' to base special relativity comparisons between inertial frames upon. It does not say whose clock is running faster or slower than anybody elses, but it does say unequivocally who will measure the shorter time and distance between two events — it the observer who is present at both events.

The special theory of relativity is a modern physical theory due to Einstein, giving an account in which neither space nor time has an independent absolute value or existence but is each relative to the other. Thus the classical view of space and time is replaced with one in which the two are aspects of the same underlying reality: Space-time. The general theory of relativity extended the special (from considering frameworks in uniform relative motion to considering frameworks in arbitrary relative motion to one another) and is the currently accepted basis for our theory of gravitation

Newton's laws of motion give us a complete description of the behaviour moving objects at low speeds. The laws are different at speeds reached by the particles at SLAC.

Einstein's Special Theory of Relativity describes the motion of particles moving at close to the speed of light. In fact, it gives the correct laws of motion for any particle. This doesn't mean Newton was wrong, his equations are contained within the relativistic equations. Newton's "laws" provide a very good approximate form, valid when v is much less than c. For particles moving at slow speeds (very much less than the speed of light), the differences between Einstein's laws of motion and those derived by Newton are tiny. That's why relativity doesn't play a large role in everyday life. Einstein's theory supercedes Newton's, but Newton's theory provides a very good approximation for objects moving at everyday speeds.

Einstein's theory is now very well established as the correct description of motion of relativistic objects, that is those traveling at a significant fraction of the speed of light.

Because most of us have little experience with objects moving at speeds near the speed of light, Einstein's predictions may seem strange. However, many years of high energy physics experiments have thoroughly tested Einstein's theory and shown that it fits all results to date.

THEORETICAL BASIS FOR SPECIAL RELATIVITY

Einstein's theory of special relativity results from two statements – the two basic postulates of special relativity:

1. The speed of light is the same for all observers, no matter what their relative speeds.
2. The laws of physics are the same in any inertial (that is, non-accelerated) frame of reference. This means that the laws of physics observed by a hypothetical observer traveling with a relativistic particle must be the same as those observed by an observer who is stationary in the laboratory.

Given these two statements, Einstein showed how definitions of momentum and energy must be refined and how quantities such as length and time must change from one observer to another in order to get consistent results for physical quantities such as particle half-life. To decide whether his postulates are a correct theory of nature, physicists test whether the predictions of Einstein's theory match observations. Indeed many such tests have been made – and the answers Einstein gave are right every time!

THE SPEED OF LIGHT IS THE SAME FOR ALL OBSERVERS

The first postulate – the speed of light will be seen to be the same relative to any observer, independent of the motion of the observer – is the crucial idea that led Einstein to formulate his theory. It means we can define a quantity c, the speed of light, which is a fundamental constant of nature.

Note that this is quite different from the motion of ordinary, massive objects. If I am driving down the freeway at 50 miles per hour relative to the road, a car traveling in the same direction at 55 mph has a speed of only 5 mph relative to me, while a car coming in the opposite direction at 55 mph approaches me at a rate of 105 mph. Their speed relative to me depends on my motion as well as on theirs.

PHYSICS IS THE SAME FOR ALL INERTIAL OBSERVERS

This second postulate is really a basic though unspoken assumption in all of science – the idea that we can formulate rules of nature which do not depend on our particular observing situation. This does not mean that things behave in the same way on the earth and in space, e.g. an observer at the surface of the earth is affected by the earth's gravity, but it does mean that the effect of a force on an object is the same independent of what causes the force and also of where the object is or what its speed is.

Einstein developed a theory of motion that could consistently contain both the same speed of light for any observer and the familiar addition of velocities described above for slow-moving objects. This is called the special theory of relativity, since it deals with the relative motions of objects.

Note that Einstein's General Theory of Relativity is a separate theory about a very different topic – the effects of gravity.

RELATIVISTIC DEFINITIONS

Physicists call particles with v/c comparable to 1 "relativistic" particles. Particles with $v/c << 1$ (very much less than one) are "non-relativistic." At SLAC, we are almost always dealing with relativistic particles. Below we catalogue some essential differences between the relativistic quantities the more familiar non-relativistic or low-speed approximate definitions and behaviours.

GAMMA()

The measurable effects of relativity are based on gamma. Gamma depends only on the speed of a particle and is always larger than 1. By definition:

For example, when an electron has traveled ten feet along the accelerator it has a speed of 0.99c, and the value of gamma at that speed is 7.09. When the electron reaches the end of the linac, its speed is 0.99999999995c where gamma equals 100,000.

What do these gamma values tell us about the relativistic effects detected at SLAC? Notice that when the speed of the object is very much less than the speed of light ($v \ll c$), gamma is approximately equal to 1. This is a non-relativistic situation (Newtonian).

MOMENTUM

For non-relativistic objects Newton defined momentum, given the symbol p, as the product of mass and velocity — $p = m\,v$. When speed becomes relativistic, we have to modify this definition — $p = \text{gamma}\,(mv)$

Notice that this equation tells you that for any particle with a non-zero mass, the momentum gets larger and larger as the speed gets closer to the speed of light. Such a particle would have infinite momentum if it could reach the speed of light. Since it would take an infinite amount of force (or a finite force acting over an infinite amount of time) to accelerate a particle to infinite momentum, we are forced to conclude that a massive particle always travels at speeds less than the speed of light.

Some text books will introduce the definition m_0 for the mass of an object at rest, calling this the "rest mass" and define the quantity ($M = \text{gamma}\ m_0$) as the mass of the moving object. This makes Newton's definition of momentum still true provided you choose the correct mass. In particle physics, when we talk about mass we always mean mass of an object at rest and we write it as m and keep the factor of gamma explicit in the equations.

ENERGY

Probably the most famous scientific equation of all time, first derived by Einstein is the relationship $E = mc^2$.

This tells us the energy corresponding to a mass m at rest. What this means is that when mass disappears, for example in a nuclear fission process, this amount of energy must appear in some other form. It also tells us the total energy of a particle of mass m sitting at rest.

Einstein also showed that the correct relativistic expression for the energy of a particle of mass m with momentum p is $E^2 = m^2c^4 + p^2c^2$. This is a key equation for any real particle, giving the relationship between its energy (E), momentum (p), and its rest mass (m).

If we substitute the equation for p into the equation for E above, with a little algebra, we get E = gamma mc^2, so energy is gamma times rest energy. (Notice again that if we call the quantity M =gamma m the mass of the particle then $E = Mc^2$ applies for any particle, but remember, particle physicists don't do that.)

Let's do a calculation. The rest energy of an electron is 0.511 MeV. As we saw earlier, when an electron has gone about 10 feet along the SLAC linac, it has a speed of 0.99c and a gamma of 7.09. Therefore, using the equation E = gamma x the rest energy, we can see that the electron's energy after ten feet of travel is 7.09 x 0.511 MeV = 3.62 MeV. At the end of the linac, where gamma = 100,000, the energy of the electron is 100,000 x 0.511 MeV = 51.1 GeV.

The energy E is the total energy of a freely moving particle. We can define it to be the rest energy plus kinetic energy ($E = KE + mc^2$) which then defines a relativistic form for kinetic energy. Just as the equation for momentum has to be altered, so does the low-speed equation for kinetic energy ($KE = (1/2)mv^2$). Let's make a guess based on what we saw for momentum and energy and say that relativistically KE = gamma$(1/2)mv^2$. A good guess, perhaps, but it's wrong.

Now here is an exercise for the interested reader. Calculate the quantity $KE = E - mc^2$ for the case of v very much smaller than c, and show that it is the usual expression for kinetic energy ($1/2\ mv^2$) plus corrections that are proportional to $(v/c)^2$ and higher powers of (v/c). The complicated result of this exercise points out why it is not useful to separate the energy of a relativistic particle into a sum of two terms, so when particle physicists say "the energy of a moving particle" they mean the total energy, not the kinetic energy.

Another interesting fact about the expression that relates E and p above ($E^2 = m^2c^4 + p^2c^2$), is that it is also true for the case where a particle has no mass (m=0). In this case, the particle always travels at a speed c, the speed of light. You can regard this equation as a definition of momentum for such a mass-less particle. Photons have kinetic energy and momentum, but no mass!

In fact Einstein's relationship tells us more, it says Energy and mass are interchangeable. Or, better said, rest mass is just one form of energy. For a compound object, the mass of the composite is not just the sum of the masses of the constituents but the sum of their energies, including kinetic, potential, and mass energy. The equation $E=mc^2$ shows how to convert between energy units and mass units. Even a small mass corresponds to a significant amount of energy.

- In the case of an atomic explosion, mass energy is released as kinetic energy of the resulting material, which has slightly less mass than the original material.
- In any particle decay process, some of the initial mass energy becomes kinetic energy of the products.

Even in chemical processes there are tiny changes in mass which correspond to the energy released or absorbed in a process. When chemists talk about conservation of mass, they mean that the sum of the masses of the atoms involved does not change. However, the masses of molecules are slightly smaller than the sum of the masses of the atoms they contain (which is why molecules do not just fall apart into atoms). If we look at the actual molecular masses, we find tiny mass changes do occur in any chemical reaction.

At SLAC, and in any particle physics facility, we also see the reverse effect — energy producing new matter. In the presence of charged particles a photon (which only has kinetic energy) can change into a massive particle and its matching massive antiparticle. The extra charged particle has to be there to absorb a little energy and more momentum, otherwise such a process could not conserve both energy and momentum. This process is one more confirmation of Einstein's special theory of relativity. It also is the process by which antimatter (for example the positrons accelerated at SLAC) is produced.

UNITS OF MASS, ENERGY, AND MOMENTUM

Instead of using kilograms to measure mass, physicists use a unit of energy — the electron volt. It is the energy gained by one electron when it moves through a potential difference of one volt. By definition, one electron volt (eV) is equivalent to 1.6×10^{-19} joules.

Lets look at an example of how this energy unit works. The rest mass of an electron is 9.11×10^{-31} kg. Using $E = mc^2$ and a calculator we get:

$$E = 9.11 \times 10^{-31} \text{ kg} \times (3 \times 10^{8} \text{ m/s})^2 = 8.199 \times 10^{-14} \text{ joules}$$

This gives us the energy equivalent of one electron. So, whether we say we have 9.11×10^{-31} kg or 8.199×10^{-14} joules, we really talking about the same thing — an electron. Physicists go one stage further and convert the joules to electron volts. This gives the mass of an electron as 0.511 MeV (about half a million eV).

So if you ask a high energy physicist what the mass of an electron is, you'll be told the answer in units of energy. You can blame Einstein for that!

Eagle-eyed readers will notice that if you solve $E=mc^2$ for m, you get $m=E/c^2$, so the unit of energy should be eV/c^2. What happened to the c^2? It's very simple, particle physicists choose units of length so that the speed of light = 1! How can we do that? Quite easily, as long as everyone understands the system. All we have to do is use a conversion factor to get back the "real" (i.e. everyday) units, if we want them.

Not only are mass and energy measured in eV, so is momentum. It makes life so much easier than dividing by c^2 or c all the time.

There is more information available on units in relativistic physics.

PECULIAR RELATIVISTIC EFFECTS

LENGTH CONTRACTION AND TIME DILATION

One of the strangest parts of special relativity is the conclusion that two observers who are moving relative to one another, will get different measurements of the length of a particular object or the time that passes between two events.

Consider two observers, each in a space-ship laboratory containing clocks and metre sticks. The space ships are moving relative to each other at a speed close to the speed of light. Using Einstein's theory:

- Each observer will see the metre stick of the other as shorter than their own, by the same factor gamma (-defined above). This is called length contraction.
- Each observer will see the clocks in the other laboratory as ticking more slowly than the clocks in his/her own, by a factor gamma. This is called time dilation.

In particle accelerators, particles are moving very close to the speed of light where the length and time effects are large. This has allowed us to clearly verify that length contraction and time dilation do occur.

TIME DILATION FOR PARTICLES

Particle processes have an intrinsic clock that determines the half-life of a decay process. However, the rate at which the clock ticks in a moving frame, as observed by a static observer, is slower than the rate of a static clock. Therefore, the half-life of a moving particles appears, to the static observer, to be increased by the factor gamma.

For example, let's look at a particle sometimes created at SLAC known as a tau. In the frame of reference where the tau particle is at rest, its lifetime is known to be approximately 3.05×10^{-13} s. To calculate how far it travels before decaying, we could try to use the familiar equation distance equals speed times time. It travels so close to the speed of light that we can use $c = 3 \times 10^8$ m/sec for the speed of the particle. (As we will see below, the speed of light in a vacuum is the highest speed attainable.) If you do the calculation you find the distance traveled should be 9.15×10^{-5} metres.

$$d = v\,t$$

$$d = (3 \times 10^8 \text{ m/sec})(3.05 \times 10^{-13} \text{ s}) = 9.15 \times 10^{-5} \text{ m}$$

Here comes the weird part - we measure the tau particle to travel further than this!

Pause to think about that for a moment. This result is totally contradictory to everyday experience. If you are not puzzled by it, either you already know all about relativity or you have not been reading carefully.

What is the resolution of this apparent paradox? The answer lies in time dilation. In our laboratory, the tau particle is moving. The decay time of the tau can be seen as a moving clock. According to relativity, moving clocks tick more slowly than static clocks.

We use this fact to multiply the time of travel in the taus moving frame by gamma, this gives the time that we will measure. Then this time times c, the approximate speed of the tau, will give us the distance we expect a high energy tau to travel.

What is gamma in this case? It depends on the tau's energy. A typical SLAC tau particle has a gamma = 20. Therefore, we detect the tau to decay in an average distance of $20 \times (9.15 \times 10^{-5} \text{ m}) = 1.8 \times 10^{-3}$ m or approximately 1.8 millimeters. This is 20 times further than we expect it to go if we use classical rather than relativistic physics. (Of course, we actually observe a spread of decay times according to the exponential decay law and a corresponding spread of distances. In fact, we use the measured distribution of distances to find the tau half-life.)

Observations particles with a variety of velocities have shown that time dilation is a real effect. In fact the only reason cosmic ray muons ever reach the surface of the earth before decaying is the time dilation effect.

LENGTH CONTRACTION

Instead of analyzing the motion of the tau from our frame of reference, we could ask what the tau would see in its reference frame. Its half-life in its reference frame is 3.05×10^{-13} s. This does not change. The tau goes nowhere in this frame.

How far would an observer, sitting in the tau rest frame, see an observer in our laboratory frame move while the tau lives?

We just calculated that the tau would travel 1.8 mm in our frame of reference. Surely we would expect the observer in the tau frame to see us move the same distance relative to the tau particle. Not so says the tau-frame observer – you only moved 1.8 mm/gamma = 0.09 mm relative to me. This is length contraction.

How long did the tau particle live according to the observer in the tau frame? We can rearrange d = v x t to read t = d/v. Here we use the same speed, Because the speed of the observer in the lab relative to the tau is just equal to (but in the opposite direction) of the speed of the tau relative to the observer in the lab, so we can use the same speed. So time = 0.09×10^{-3} m/(3×10^{8})m/sec = 3.0×10^{-13} sec. This is the half-life of the tau as seen in its rest frame, just as it should be!

OTHER IMPLICATIONS OF SPECIAL RELATIVITY

$E = mc^2$

One thing you will notice about the procedure defined at the end of the previous section is that the gain when you try to accelerate a ball to the speed of light in steps, the increase in velocity relative to the initial frame gets less and less, each time you repeat the process. The faster the ball is going, the hard it is to accelerate further. Now recall the definition of inertial mass: it is the measure of an object's resistance to a change in motion. In Newtonian physics inertial mass is a property of the object alone, and does not depend in any way on its state of motion. A one kilogram ball has a mass of one kilogram no matter how fast it happens to be moving. The situation is drastically different in Special Relativity. The faster a mass moves, the harder it is to accelerate. Hence its inertial mass depends on its speed! It will come as no surprise that this dependence is determined by the time dilation factor . In particular, if an object is moving relative to you at a speed of v, you will measure an inertial mass that increases by a factor of compare to its value m_0 at rest:

m_0 is called the rest mass of the object, for obvious reasons. For example, if an object is moving at .99c, its mass will appear to be a factor of seven times larger than its rest mass. Again, because of the form of this effect is negligible at velocities that are small compared to the speed of light. However, the increase of inertial mass with velocity has been measured in the large accelerators used by particle physicists to create fast moving beams of electrons. The faster the electrons are moving, the harder it is to further increase their kinetic energy.

There is another profound implication of the above considerations that Einstein was the first to realize. We know that when you increase the velocity of an object, you are giving it kinetic energy. We also know that kinetic energy can be converted into other forms of energy, such as potential energy and heat energy and vice versa. If increasing the velocity of an object increases its inertial mass, then in some sense we are converting kinetic energy into mass. Einstein reasoned that mass and energy must somehow be different manifestations of the same thing. In particular if we can convert kinetic energy into mass, we should also be able to convert mass into energy. The special theory of relativity in fact gives an explicit expression for how mass and energy are related. You have probably seen it before.

Thus as expected from earlier objects, the energy of an object increase without bound as its velocity approaches the speed of light. The surprising aspect of equation is that even when an object is at rest, it has energy. When v = 0, = 1 amd m = m_0 so that for a mass at rest:

The energy of an object at rest is related to its mass by the factor c^2 which is a very large number. The consequence is that very small masses can produce vast amounts of energy, and nuclear power plants and bombs are effective because they can convert mass into other forms of energy. In fact, if all the mass in a single grain of sand could be converted into energy, it could provide enough electricity to light up a whole city for a year.

SPACE TRAVEL

The velocity addition law implies that no object can be accelerated to speeds greater than the speed of light. At first glance this seems to put a damper on the prospects for space travel beyond our solar system. The nearest star to our Sun is four light years away, and to get to a potentially more interesting star, say one with planets, you would have to travel hundreds of light years. A light year is the unit of distance corresponding to the distance light travels in a year. It is a terribly long distance, about 10^{12}kilometers. More importantly, the speed limit imposed by Special Relativity forbids us from getting anywhere in less time than it would take light to get there, so that trips to even nearby stars would take many years, and perhaps even hundreds of years. If this were true it would make travel between the stars undesirable at best for a species whose life span is only about one hundred years. Luckily, what Special Relativity takes away with one hand, it gives back with the other.

A trip to the nearest star would take approximately four years for a spaceship traveling at 99% of the speed of light. However, this is the time as measure by observers stationary with respect to the Earth and the star (we assume that the relative motion between them is nowhere near light speed). According to the time dilation effect, the time required for the same journey would be shorter to those on the spaceship. For example, if the spaceship is moving at 0.99c, the time dilation factor is seven, so that to the astronauts, the journey takes only 4/7 of a year, or about seven months. If the spaceship were able to reach 0.999c, the time dilation factor would be twenty, and the trip would take only about one month. Space travel, and the colonization of space are therefore possible despite the relativistic speed limit. The only downside is that the taxpayers on Earth who funded the mission would still have to wait at least eight years before the astronauts returned, even though the astronauts themselves would have only aged two months.

THE TWIN PARADOX

The previous subsection illustrates a curious feature of Special Relativity that, at first glance seems paradoxical. We argued that astronauts could travel to a star four light years from Earth and only age two months if they traveled at 0.999c. Meanwhile the people back on Earth age the full eight years. This seems to contradict one of the basic postulates of Special Relativity, namely the relativity of uniform motion. After all, time dilation should be symmetrical, so how is it that the astronauts age less during the journey instead of the people on Earth. In fact the prediction that the astronauts age less is the correct one. It has been experimentally verified, on a smaller scale, but putting very accurate atomic clocks in orbit around the Earth, and comparing them to their Earthbound counterparts after the journey. The clocks that went into orbit were slower in accordance with the predictions of Special Relativity.

This dilemma highlights a limitation of the Special Theory of Relativity that we have already alluded to. It only applies to observers in uniform motion, and not to accelerated frames. In order for astronauts to go to a distance star, and return to compare clocks with Earthbound observers, the astronauts' spaceship must accelerate to near light speeds, decelerate once the reach the star, and the repeat the process in the other direction. While they are accelerating, the rules of Special

Relativity don't apply, and the symmetry between the astronauts and the Earthbound observers breaks down. (This applies to the atomic clocks in orbit as well.) A detailed examination of the problem therefore requires us to go beyond Special Relativity (to General Relativity), with the result that time actually slows down in accelerated frames of reference. However, it turns out that the calculation we did in the previous sub-section is approximately correct. We get nearly the right answer for the difference in aging of the astronauts by ignoring the acceleration and just taking into account time dilation on the forward and return journeys. Thus, there is no "paradox" associated with this effect. As with most paradoxes, it merely points to a limitation of the theory under consideration.

TACHYONS AND TIME TRAVEL

Special Relativity states that nothing that is moving initially slower than the speed of light can be accelerated to exceed the speed of light. This is not strictly speaking the same as the statement that nothing can travel faster than light. It is in principle possible that particles exist which move faster than light from the moment they are created. If such particles did exist, however, Special Relativity implies that they could never be slowed down to velocities below the speed of light. The existence of such particles, called tachyons, has not been totally ruled out, but several experiments have tried, without luck to detect them. If they did exist, and they interacted with ordinary matter, it would give us the means to communicate with the past. Tachyons could literally be sent outwards, bounce off a tachyonic mirror, and return before they were sent. This in turn would give rise to a great many logical problems. For example, if you sent a message back in time that caused your grandfather to be killed before your father was conceived, then you would not be around to send the message that prevented your birth, so that you would be around to send the message, so that you wouldn't be around....etc.

THE APPARENT INCOMPATIBILITY OF THE LAW OF PROPAGATION OF LIGHT WITH THE PRINCIPLE OF RELATIVITY

There is hardly a simpler law in physics than that according to which light is propagated in empty space. Every child at school knows, or believes he knows, that this propagation takes place in straight lines with a velocity c = 300,000 km./sec. At all events we know with great exactness that this velocity is the same for all colours, because if this were not the case, the minimum of emission would not be observed simultaneously for different colours during the eclipse of a fixed star by its dark neighbour. By means of similar considerations based on observations of double stars, the Dutch astronomer De Sitter was also able to show that the velocity of propagation of light cannot depend on the velocity of motion of the body emitting the light. The assumption that this velocity of propagation is dependent on the direction "in space" is in itself improbable.,

In short, let us assume that the simple law of the constancy of the velocity of light c (in vacuum) is justifiably believed by the child at school. Who would imagine that this simple law has plunged the conscientiously thoughtful physicist into the greatest intellectual difficulties? Let us consider how these difficulties arise.,

Of course we must refer the process of the propagation of light (and indeed every other process) to a rigid reference-body (co-ordinate system). As such a system let us again choose our embankment. We shall imagine the air above it to have been removed. If a ray of light be sent along the embankment, we see from the above that the tip of the ray will be transmitted with the velocity c relative to the embankment. Now let us suppose that our railway carriage is again travelling along the railway lines with the velocity v, and that its direction is the same as that of the ray of light, but its velocity of course much less. Let us inquire about the velocity of propagation of the ray of light relative to the carriage. It is obvious that we can here apply the consideration of the previous section, since the ray of light plays the part of the man walking along relatively to the carriage. The velocity W of the man relative to the embankment is here replaced by the velocity of light relative to the embankment. w is the required velocity of light with respect to the carriage, and we have

$$w = c - v.$$

The velocity of propagation of a ray of light relative to the carriage thus comes out smaller than c.,

But this result comes into conflict with the principle of relativity set forth in early section. For, like every other general law of nature, the law of the transmission of light in vacuo must, according to the principle of relativity, be the same for the railway carriage as reference-body as when the rails are the body of reference. But, from our above consideration, this would appear to be impossible. If every ray of light is propagated relative to the embankment with the velocity c, then for this reason it would appear that another law of propagation of light must necessarily hold with respect to the carriage—a result contradictory to the principle of relativity.

In view of this dilemma there appears to be nothing else for it than to abandon either the principle of relativity or the simple law of the propagation of light in vacuo. Those of you who have carefully followed the preceding discussion are almost sure to expect that we should retain the principle of relativity, which appeals so convincingly to the intellect because it is so natural and simple. The law of the propagation of light in vacuo would then have to be replaced by a more complicated law conformable to the principle of relativity. The development of theoretical physics shows, however, that we cannot pursue this course. The epoch-making theoretical investigations of H. A. Lorentz on the electrodynamical and optical phenomena connected with moving bodies show that experience in this domain leads conclusively to a theory of electromagnetic phenomena, of which the law of the constancy of the velocity of light in vacuo is a necessary consequence.

Prominent theoretical physicists were therefore more inclined to reject the principle of relativity, in spite of the fact that no empirical data had been found which were contradictory to this principle.,

At this juncture the theory of relativity entered the arena. As a result of an analysis of the physical conceptions of time and space, it became evident that in reality there is not the least incompatibility between the principle of relativity and the law of propagation of light, and that by systematically holding fast to both these laws a logically rigid theory could be arrived at. This theory has been called the special theory of relativity to distinguish it from the extended theory, with which we shall deal later. In the following pages we shall present the fundamental ideas of the special theory of relativity.

CAN SPECIAL RELATIVITY HANDLE ACCELERATIONS

It is a common misconception that Special Relativity cannot handle accelerating objects or accelerating reference frames. It is claimed that general relativity is required because special relativity only applies to inertial frames. This is not true. Special relativity treats accelerating frames differently from inertial frames but can still deal with them. Accelerating objects can be dealt with without even calling upon accelerating frames.

This error often comes up in the context of the twin paradox when people claim that it can only be resolved in general relativity because of acceleration. This is not the case.

The only sense in which special relativity is an approximation when there are accelerating bodies is that gravitational effects such as generation of gravitational waves are being ignored. But of course there are larger gravitational effects being neglected even when massive bodies are not accelerating and they are small for many applications so this is not strictly relevant. Special relativity gives a completely self-consistent description of the mechanics of accelerating bodies neglecting gravitation, just as Newtonian mechanics did.

The difference between general and special relativity is that in the general theory all frames of reference including spinning and accelerating frames are treated on an equal footing. In special relativity accelerating frames are different from inertial frames. Velocities are relative but acceleration is treated as absolute. In general relativity all motion is relative. To accommodate this change general relativity has to use curved space-time. In special relativity space-time is always flat.

In special relativity an accelerating particle has a worldline which is not straight. This is not difficult to handle. The 4-vector acceleration can be defined as the derivative with respect to proper time of the 4-velocity. It is possible to solve the equations of motion for a particle in electric and magnetic fields, for example.

Accelerating reference frames are a different matter. In GR the physical equations take the same form in any co-ordinate system. In SR they do not but it is still possible to use co-ordinate systems corresponding to accelerating or rotating frames of reference just as it is possible to solve ordinary mechanics problems in curvilinear co-ordinate systems. This is done by introducing a metric tensor. The formalism is very similar to that of many general relativity problems but it is still special relativity so long as the space-time is constrained to be flat and Minkowskian. Note that the speed of light is rarely constant in non-inertial frames and this has been known to cause confusion.

An example would be a rotating frame of reference used to deal with a rotating object. The transformation of the metric into the rotating frame would lead to "fictitious" forces such as Coriolis forces and centrifugal forces. It is not very different from ordinary mechanics.

A simple problem is to solve the motion of a body which accelerates constantly. What does this mean? We don't mean that its acceleration as measured by an inertial observer is constant. We mean that it is moving so that the acceleration measured in an inertial frame travelling at the same instantaneous velocity as the object is the same at any moment. If it was a rocket and you were on board you would experience a constant G force. This problem can be solved in a number of ways. One is to use four-vector acceleration along its worldline which must have constant magnitude. Alternatively, the object is passing constantly from one inertial frame to another in such a way that its change of speed in a fixed time interval seen as a Lorentz boost is always the same. From our understanding of adding velocities we can see that the rapidity r of the object must be increasing at a constant rate a with respect to the proper time of the object T. The rapidity is related to velocity v by the equation

$$v = c \tanh(r/c)$$

From this we derive the equation

$$v = c \tanh(aT/c)$$

Chapter 9

General Relativity

Einstein's general relativity is mostly about gravitation, curved space-time, geodesics and black holes, the latter being the de facto trademark of the theme.

In this section there are many topic you need to know prior to understand this whole concept.

There is no simple answer to the age-old question what is gravity? It seems like a mysterious force that attracts objects towards each other. The force seems not to care from which materials the objects are made. What is more, the force apparently operates instantly over large distances. Even more puzzling is the fact that no object can be shielded from the force of gravity.

Newton discovered that this mysterious force is directly proportional to the product of the masses of the objects and inversely proportional to the square of the distance between them. This force will cause any two objects in free space to fall towards each other at a predictable acceleration and speed.

This acceleration is, strangely enough, independent of the mass of the object for which you measure the acceleration. This is so, because in Newton's theory, gravitational mass is equal to inertial mass. Your gravitational mass is roughly what your bathroom scale says. Your inertial mass is, loosely speaking, the resistance of your body against being pushed around.

In the Newtonian calculation of the gravitational acceleration of a body, the inertial mass and the gravitational mass, which are equal, cancel out. In a vacuum environment, this causes all bodies to fall with the same acceleration towards a large mass, like Earth.

Einstein abandoned the notion of gravity as a force and explained it as caused by objects attempting to follow straight lines in curved space-time.

Gravitational Forces

Newton eventually came to the conclusion that, in fact, the apple and the moon were influenced by the same force. He named that force gravitation (or gravity) after the Latin word gravitas which literally translates into "heaviness" or "weight."

In the Principia, Newton defined the force of gravity in the following way (translated from the Latin):

Every particle of matter in the universe attracts every other particle with a force that is directly proportional to the product of the masses of the particles and inversely proportional to the square of the distance between them.

Mathematically, this translates into the force equation shown to the right. In this equation, the quantities are defined as:

- F_g = The force of gravity (typically in newtons)
- G = The gravitational constant, which adds the proper level of proportionality to the equation. The value of G is 6.67259×10^{-11} N * m^2 / kg^2, although the value will change if other units are being used.
- m_1 & m_1 = The masses of the two particles (typically in kilograms)
- r = The straight-line distance between the two particles (typically in metres)

Interpreting the Equation

This equation gives us the magnitude of the force, which is an attractive force and therefore always directed toward the other particle. As per Newton's Third Law of Motion, this force is always equal and opposite.

Newton's Three Laws of Motion give us the tools to interpret the motion caused by the force and we see that the particle with less mass (which may or may not be the smaller particle, depending upon their densities) will accelerate more than the other particle. This is why light objects fall to the Earth considerably faster than the Earth falls toward them. Still, the force acting on the light object and the Earth is of identical magnitude, even though it doesn't look that way.

It is also significant to note that the force is inversely proportional to the square of the distance between the objects. As objects get further apart, the force of gravity drops very quickly. At most distances, only objects with very high masses such as planets, stars, galaxies, and black holes have any significant gravity effects.

Centre of Gravity

In an object composed of many particles, every particle interacts with every particle of the other object. Since we know that forces (including gravity) are vector quantities, we can view these forces as having components in the parallel and perpendicular directions of the two objects. In some objects, such as spheres of uniform density, the perpendicular components of force will cancel each other out, so we can treat the objects as if they were point particles, concerning ourselves with only the net force between them.

The centre of gravity of an object (which is generally identical to its centre of mass) is useful in these situations. We view gravity, and perform calculations, as if the entire mass of the object were focused at the centre of gravity. In simple shapes - spheres, circular disks, rectangular plates, cubes, etc. - this point is at the geometric centre of the object.

This idealized model of gravitational interaction can be applied in most practical applications, although in some more esoteric situations such as a non-uniform gravitational field, further care may be necessary for the sake of precision.

WHAT IS QUANTUM GRAVITY?

Quantum gravity is an overall term for theories that attempt to unify gravity with the other fundamental forces of physics (which are already unified together). It generally posits a theoretical entity, a graviton, which is a virtual particle that mediates the gravitational force. This is what distinguishes quantum gravity from certain other unified field theories ... although, in fairness, some theories that are typically classified as quantum gravity don't necessary require a graviton.

What's a Graviton?

The standard model of quantum mechanics (developed between 1970 & 1973) postulates that the other three fundamental forces of physics are mediated by virtual bosons.

Photons mediate the electromagnetic force, W & Z bosons mediate the weak nuclear force, and gluons (such as quarks) mediate the strong nuclear force.

The graviton, therefore, would mediate the gravitational force. If found, the graviton is expected to be massless (because it acts instantaneously at long distances) and have spin 2 (because gravity is a second-rank tensor field).

Is Quantum Gravity Proven?

The major problem in experimentally testing any theory of quantum gravity is that the energy levels required to observe the conjectures are unattainable in current laboratory experiments.

Even theoretically, quantum gravity runs into serious problems. Gravitation is currently explained through the theory of general relativity, which makes very different assumptions about the universe at the macroscopic scale than those made by quantum mechanics at the microscopic scale.

Attempts to combine them generally run into the "renormalization problem," in which the sum of all of the forces do not cancel out and result in an infinite value. In quantum electrodynamics, this happened occasionally, but one could renormalize the mathematics to remove these issues. Such renormalization does not work in a quantum interpretation of gravity.

The assumptions of quantum gravity are generally that such a theory will prove to be both simple and elegant, so many physicists attempt to work backward, predicting a theory that they feel might account for the symmetries observed in current physics and then seeing if those theories work.

Some unified field theories that are classified as quantum gravity theories include:

- String theory / Superstring theory / M-theory
- Supergravity
- Loop quantum gravity
- Twistor theory
- Noncommutative geometry
- Euclidean quantum gravity
- Wheeler-deWitt equation

Of course, it's fully possible that if quantum gravity does exist, it will be neither simple nor elegant, in which case these attempts are being approached with faulty assumptions and, likely, would be inaccurate. Only time, and experimentation, will tell for sure.

It is also possible, as some of the above theories predict, that an understanding of quantum gravity will not merely consolidate the theories, but will rather introduce a fundamentally new understanding of space and time

SURPRISING VELOCITY EFFECTS ON GRAVITY AND ACCELERATION

The fact that there are velocity effects on gravity may not come as a surprise now that you know that space-time curvature is the root cause of gravity. When you view objects moving linearly on a curved surface, you will expect to observe non-linear motion, depending from where you view it.

The effects of movement on gravity can be expressed as part of the gravitational acceleration equations.

How Orbits Work

Most engineers know how orbits work, especially Kepler's laws of planetary motion. Some engineers also know that Newton's laws of motion and of gravity can be used to plot orbits. Relativistic orbits are however somewhat "mysterious". This page and its download will show you all the basics of orbits, both for Newtonian and relativistic situations.

Relativistic orbital situations normally arise when the closest point or periapsis of an orbit comes close to a very massive body. More specifically, it is when the fraction: Newton's gravitational constant (G) times the primary mass, divided by (the periapsis distance times the square of the speed of light) becomes larger than roughly 1 in 100 million. All values are in SI units (kg, m, sec).

In the case of our solar system, it is only planet Mercury that enters a (slightly) relativistic scenario, where the above "relativistic fraction" is about one in 30 million - still very tiny. It is however enough to make Mercury's perihelion shift by an additional 42 arc-seconds per century. This is over and above what can be predicted by using Newton's theory

ORBITAL EQUATIONS IN EINSTEIN'S RELATIVITY

Kepler's laws for planetary motion teaches us that the orbital equations for closed orbits must be elliptical in nature. More generally, orbits are conic sections, i.e., circular, elliptical, parabolical or hyperbolical.

Newtonian orbital equations are normally obtained by writing down Newton's equation of motion for an object moving in a gravitational field. The solution to the equation of motion gives the formula for a conic section.

The orbital formula is that of a circle for an eccentricity of zero, an ellipse if the eccentricity is less than unity, a parabola for eccentricity precisely unity and a hyperbola for eccentricity greater than unity.

Relativistic orbits are much more complex. The equations are usually obtained by writing down the formulas for a space-time geodesic and then partially solving the equations. There are no exact solutions for relativistic orbits, i.e., solutions enabling direct plots of the orbits do not exist.

There are various ways to obtain approximate solutions to the equations, depending on how relativistic the orbits become. Another popular way is to solve the equations numerically, i.e., step-by-step solutions, also called "numerical integration".

TESTS OF RELATIVITY

There have been numerous tests of relativity over the more than a century since Einstein published his first paper on the subject. The most complex tests are those for small general relativistic effects, using our solar system as a "test-bench".

The three major tests of relativity touched upon in this page and its attachments are: the perihelion shift of Mercury, the deflection of light by the Sun and the Shapiro time delay of light. Then there is the very contemporary Gravity Probe B, the measurement of the geodetic effect and inertial frame dragging.

The perihelion Shift of Mercury

The anomalous perihelion shift of Mercury was a well-known fact by the time Einstein completed his general theory of relativity. After astronomers have compensated for all know effects of other planets on the orbit of Mercury, they were left with an inexplicable 43 arc-seconds per century.

Einstein's first "reality check" on his theory of general relativity was indeed to use an approximate solution to his field equations and calculate the perihelion shift of Mercury. It came out so close to the 43 arc-seconds per century that Einstein said it gave him "palpitations of the heart".

Deflection of Light by Gravity

Einstein also predicted that the Sun would deflect a close-passing (or "grazing") ray of light by about 1.75 arc-seconds.

Sir Arthur Eddington measured the deflection shortly after WWII. Although his measurement was not very accurate, it apparently confirmed the deflection as predicted by Einstein. This was the beginning of Einstein's fame.

The Time Delay of Light

Light is "delayed" when it passes close by the Sun, or close by any other very massive object. Einstein did not predict this effect. It was Irvin Shapiro who first made the calculations and performed experiments to confirm the prediction. He bounced radar signals off Mars when it passed near to directly behind the Sun and measured the time.

The time it should take for the radar signals to make the return trip to Mars (without passing the Sun) is calculated using all available astronomical information, e.g., the positions of Earth and Mars relative to each other at any moment.

The difference between the calculated time and the actual measured time is the so-called Shapiro time delay of light. It is about 250 microseconds on a total round trip time of about 45 minutes.

The attached pdf file will give you an engineering view of the above tests of relativity.

Gravity Probe B

Gravity Probe B is a spacecraft in a polar orbit about 650 km above Earth, intended to measure the properties of space-time around Earth. The two effects that it is designed to measure are the "geodetic effect" and "inertial frame dragging".

The geodetic effect is caused directly by space-time curvature. A gyroscope in space 'leans over' into the slope of space-time. When moved around earth in one orbit, the spin axis will point in a slightly different direction than the direction it started from.

Inertial frame dragging or 'gravitomagnetism' is caused by any rotating mass, which tends to drag the close-by space-time around with it, somewhat like a ball spinning in liquid does with the liquid. This effect alters the spin axis of the gyroscope in a different direction than the geodetic effect (if the gyro is in a polar orbit).

TIDAL GRAVITY

Tidal gravity encompasses a lot more than the tides of the ocean. The title of this page might suggest that we are talking about tides - more specifically coastal tides, but there is more... We will spend some time on coastal tides. In fact, it will be shown that there is a common misconception about how coastal tides work.

In Newton's theory, tidal gravity is the fact that parts of extended objects are at different distances from a source of gravity. The gravitational force vectors are therefore different on different parts of the extended object. This can stretch and/ or squeeze an object.

In relativity theory, the same effect is explained by means of curved space-time. The curvature is different for different regions of extended objects. This causes the space-time geodesics to be different for each region, hence the stretch and/ or squeeze that is experienced.

Like always, the differences between Newton's and relativistic effects are slight at large distances from objects. It is only in the very near-regions that there is a marked difference.

Tidal gravity is a relatively short-range effect. This is borne out by the fact that our moon's effect on Earth is stronger than the much more massive Sun's effect. In fact tidal effects drops off with the inverse of the cube of the distance from the primary mass.

Gravitational Waves

Gravitational waves are ripples in space-time itself. It is generated when strong fields of gravity change in periodic or non-periodic ways. This causes ripples in space-time, almost like a pebble dropped in a pond will cause ripples on an otherwise smooth surface. The ripples in space-time propagate at the speed of light through space.

These waves carry energy and will rob the source or sources of energy. When the stars in a binary star system are very close together, they are theorized to radiate a lot of orbital energy as waves in space-time. This causes the orbits of the binary system to decay.

This fact led to the first, though indirect, confirmation of gravitational waves. Observations of a binary pulsar showed that the orbits actually do decay as predicted by Einstein's theory of general relativity.

Direct detection of these ripples is very difficult because they become so weak over large distances. The distance, or more precisely, the volume of space in which sources of gravitational waves have a reasonable probability to occur, is very large.

Sources of such ripples in space-time are, amongst others: supernovae, coalescing binary stars and coalescing black holes. There are a few massive detector systems built to detect such events, but up to the date of writing (early 2006), no detection has been confirmed.

The PPN Formalism of Gravitation Theory

Now we come to a matter of gravity: how can gravity be an inverse square law force, when the distance between two objects can not even be defined in Einstein's special relativity? Special relativity was constructed to satisfy Maxwell's equations, which replaced the inverse square law electrostatic force by a set of equations describing the electromagnetic field. So gravity was the only remaining action-at-a-distance inverse square law force. And gravity has a unique property; the acceleration due to gravity at a given place and time is independent of the nature of the body.

PPN Formalism is shorthand for Parameterized Post-Newtonian Formalism. The term Post-Newtonian means that it is an approximation of the full gravitational theory. It is however closer to the truth than standard Newtonian gravitational theory.

The term Parameterized means that it is set up so that by changing various parameters, competing gravitational theories can be accommodated. The best-known competing theory is the Brans-Dicke theory of gravity.

This formalism is mostly used for calculations in the solar system, where the gravitational field is not very strong and the velocities are not very relativistic. It does however take into account all the subtleties of general relativity (and more, due to the other theories it can support). Now we come to a matter of gravity: how can gravity be an inverse square law force, when the distance between two objects can not even be defined in Einstein's special relativity? Special relativity was constructed to satisfy Maxwell's equations, which replaced the inverse square law electrostatic force by a set of equations describing the electromagnetic field. So gravity was the only remaining action-at-a-distance inverse square law force. And gravity has a unique property; the acceleration due to gravity at a given place and time is independent of the nature of the body.

The space-time diagram above shows a proton and an antiproton moving under the influence of an electric field on the left and a gravitational field on the right. Gravity accelerates all objects equally. This fact was known to Newton, and tested to an accuracy of 1 part in 100 million by Eotvos. Later work by Dicke and by Braginsky has improved the accuracy of the test to 1 part in a trillion. A space mission called STEP has been proposed which will test the equality of acceleration to 1 part per quintillion. In 2000 STEP was selected for further study in NASA's SMEX programme.

Thus through any event in space-time, in any given direction, there is only one worldline corresponding to motion solely influenced by gravity. Compare this to the geometric fact that through any point, in any given direction, there is only one straight line. We are led to propose that worldlines influenced only by gravity are really straight worldlines. But how can an accelerating body have a straight worldline? It all depends on how you measure it. Suppose we plot a straight line on polar graph paper, and then make a plot of radius vs angle as shown below?

In the radius vs theta plot the straight line is curved. We have shown adjacent two lines dashed and labeled just like the proton and antiproton shown earlier to emphasize that while there are an infinite number of curved lines in the radius vs theta plot, there is only one straight line through the initial point with the initial direction.

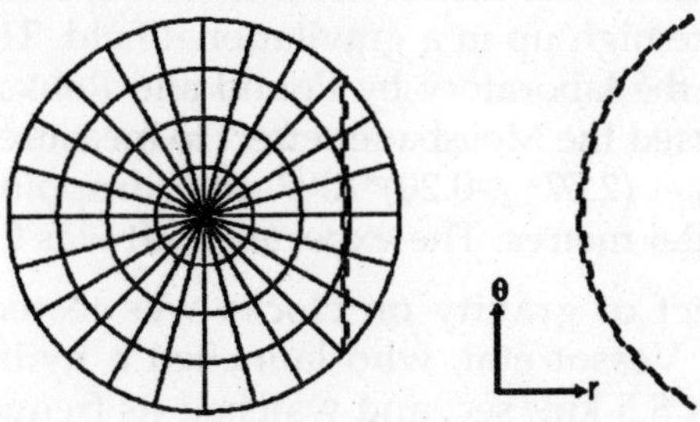

PRINCIPLE OF EQUIVALENCE

Einstein proposed that the effects of gravity (in a small region of spacetime) are equivalent to the effect of using an accelerated frame of reference without gravity. As as example, consider the famous "Einstein elevator" thought experiment. If an elevator far out in space accelerates upward at 10 metres/second^2, it will feel like a downward acceleration of gravity at 1 g = 10 m/s^2. If a clock on the ceiling of the elevator emits flashes of light f times per second, an observer on the floor will see them arriving faster than f times per second because of the Doppler shift due to the acceleration of the elevator during the light transit time.

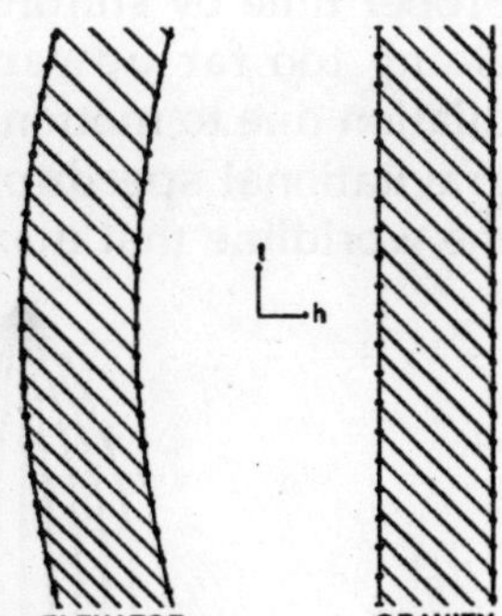

The space-time diagram on the left above shows the clock on the elevator ceiling emitting flashes of light. The light transit time is h/c where h is the height of the ceiling, and the velocity change is a*h/c so the Doppler shift increases the rate of flash arrival by a factor of $(1+a*h/c^2)$, so the flash arrival rate is $f' = (1+a*h/c^2)*f$. On the right is the same situation with stationary clocks in a gravitational field. In order to have the flash arrival rate faster by a factor of $(1+g*h/c^2)$, the clock on the ceiling must run faster by this factor. In other words, clocks run faster when they are high up in a gravitational field. This effect has been seen in the laboratory by Pound and Rebka (1960, PRL, 4, 337) who used the Mossbauer effect to measure a frequency shift $(f'/f - 1) = (2.57+/-0.20)*10^{-15}$ after dropping photons a distance of 22.6 metres. The expected shift was $2.46*10^{-15}$.

The effect of gravity on clocks was tested to greater precision by Vessot etal, who launched a hydrogen maser straight up at 8.5 km/sec, and watched its frequency change as it coasted up to 10,000 km altitude and then fell back to Earth. The frequency shift due to gravity was $(f'/f - 1) = 4*10^{-10}$ at 10,000 km altitude, and the experimental result agreed to within 70 parts per million of this shift.

Because of the gravitational speedup for uphill clocks, an observer moving between two events can achieve a larger proper time by shifting his worldline upward in the middle. Going too far upward requires moving so fast that time dilation due to motion reduces the proper time more than the gravitational speedup, so there is an optimum curvature to the worldline that maximizes the proper time.

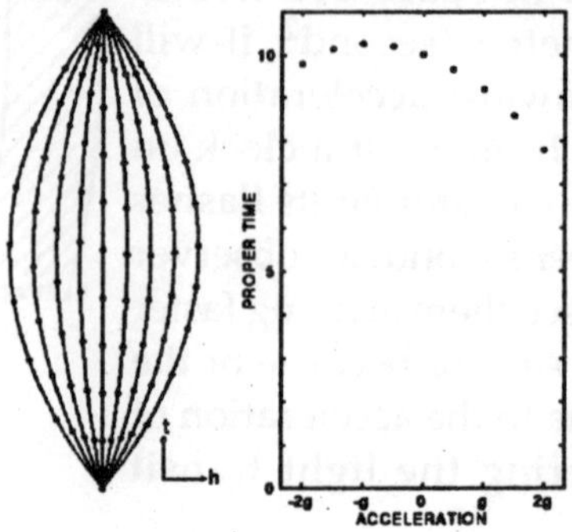

The space-time diagram on the left above shows 9 observers moving between two events with different accelerations. The third from the right has the correct balance between going uphill to get a faster clock rate and avoiding motion to avoid time dilation.

As a result, this observer has the largest proper time between the two events. Note that the accelerations are negative for paths that have a maximum height, so the third worldline from the right is plotted as the third dot from the left on the chart. The optimum worldline curvature is the acceleration of gravity, and it is negative because things fall down, not up.

CURVED SPACETIME

Curved coordinates alone, such as the polar graph, do not provide a satisfactory model for gravity. Two straight lines through the same point but with different directions will never cross again, while two worldlines influenced only by gravity which pass through the same event with different velocities can cross again. Consider the Galileo spacecraft, which made two Earth flybys.

In between the flybys, Galileo was on an elliptical orbit with a 2 year period. In order to allow "straight" lines to cross multiple times, a curved space-time is needed. As a familiar example of a curved space, consider the surface of the Earth and the great circle arc connecting two cities. The great circle is the shortest distance between two points on the surface of the Earth, and it is the path followed by airliners.

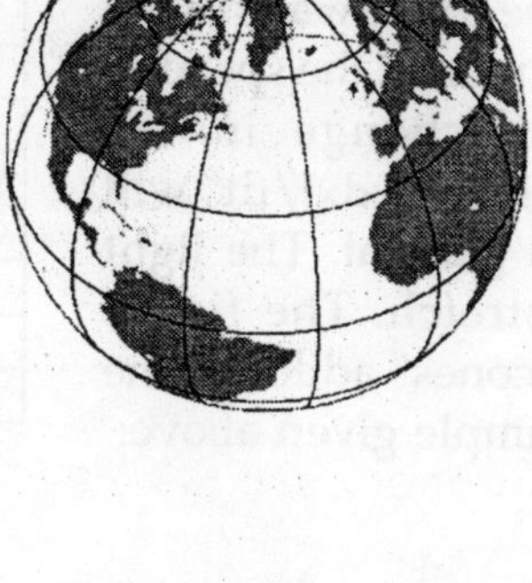

The great circle path from Los Angeles (34 N, 118 W) to Tel Aviv (32 N, 35 E) goes all the way to 70 N latitude.

Plotting latitude vs longitude, as if longitude were time and latitude position, gives the pseudo-spacetime diagram below.

The two great circles through Los Angeles, one to Tel Aviv and one to Singapore, are both "straight" lines, but they intersect in two places. This is impossible in plane geometry but it does occur in non-Euclidean geometry. The pseudo-spacetime diagram above is almost identical to a real spacetime diagram for objects moving in a tunnel drilled through the centre of a massive sphere. Gravity produces oscillatory motions so worldlines for different objects, each influenced only by gravity, can cross at many events.

Einstein was able to compute the perihelion advance of Mercury using general relativity, and his calculation matched an observed discrepancy with Newtonian predictions. Einstein also computed that light passing by the Sun would be deflected by twice as much as a prediction using Newtonian gravity and Newton's particle model for light would suggest. The same effect causes a delay of light passing by the limb of the Sun, known as the Shapiro delay, which has now been measured to great accuracy and agrees with the prediction of general relativity.

LIGHT CONES IN CURVED SPACE-TIME

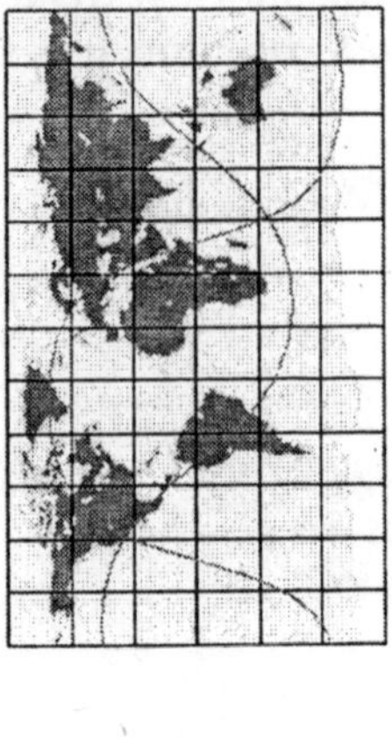

Unlike the restricted set of Lorentz transformations allowed in special relativity, the more general coordinate transformations of general relativity will change the slope of the walls of the lightcones. In other words, the speed of light (dx/dt) will change in the transformed coordinates: dx'/dt' will not equal dx/dt in general. The light cones can tilt or stretch. The figure below shows "lightcones" added to the radius vs angle example given above:

The proper generalization of the special relativistic rule that velocities of objects should be less than c is that the worldlines of objects should be inside the lightcones, but the lightcones can tilt over or stretch vertically or horizontally. It is quite possible for a stationary object to have a worldline that is outside the lightcone. This happens inside the stationary limit of a black hole, for example, or for distances greater than the Hubble radius in cosmological models, as shown below in a space-time diagram.

Thus the fundamentals of relativity that are important for cosmology are:

- The speed of light is a constant independent of the velocity of the source or the observer.
- Events that are simultaneous as seen by one observer are generally not simultaneous as seen by other observers, so there can be no absolute time.
- Each observer can define his own proper time — the time measured by a good clock moving along his worldline.
- Observers can assign times and positions to events not on their worldlines using radar observations.
- Every observer will see his clock running faster than other clocks which are moving with respect to him, and this is a mathematically consistent pattern required by the properties of radar observations.
- As a result, the unaccelerated worldline between two events will have the longest proper time of all worldlines connecting these events.
- In the presence of gravity, the worldlines of objects accelerated only by gravity have the longest proper times.
- Gravity requires that spacetime have a non-Euclidean geometry, and this curvature of spacetime must be created by matter.

Chapter 10

Theory of General Relativity

In 1916 Einstein expanded his Special Theory to include the effect of gravitation on the shape of space and the flow of time.

This theory, referred to as the General Theory of Relativity, proposed that matter causes space to curve. Embedding diagrams can help us visualize the implications of Einstein's General Theory of Relativity. Embedding Diagrams

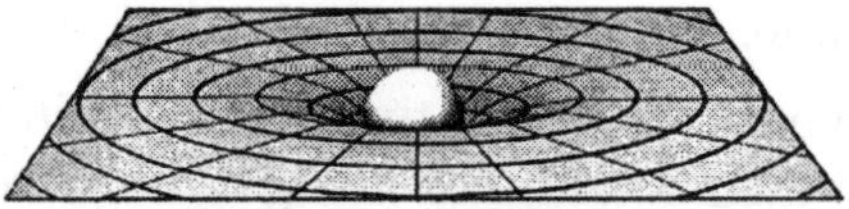

Fig. A bowling ball on a stretched rubber sheet

The large ball will cause a deformation in the sheet's surface. A baseball dropped onto the sheet will roll toward the bowling ball. Einstein theorized that smaller masses travel toward larger masses not because they are "attracted" by a mysterious force, but because the smaller objects travel through space that is warped by the larger object. Physicists illustrate this idea using embedding diagrams.

Contrary to appearances, an embedding diagram does not depict the three-dimensional "space" of our everyday experience. Rather it shows how a 2D slice through familiar 3D space is curved downwards when embedded in flattened hyperspace. We cannot fully envision this hyperspace; it contains seven dimensions, including one for time! Flattening it to 3D allows us to represent the curvature. Embedding diagrams can help us visualize the implications of Einstein's General Theory of Relativity.

THE FLOW OF SPACETIME

Another way of thinking of the curvature of spacetime was elegantly described by Hans von Baeyer. In a prize-winning essay he conceives of spacetime as an invisible stream flowing ever onward, bending in response to objects in it s path, carrying everything in the universe along its twists and turns.

This is a basic postulate of the Theory of General Relativity. It states that a uniform gravitational field (like that near the Earth) is equivalent to a uniform acceleration.

What this means, in effect, is that a person cannot tell the difference between (a) standing on the Earth, feeling the effects of gravity as a downward pull and (b) standing in a very smooth elevator that is accelerating upwards at just the right rate of exactly 32 feet per second squared.

In both cases, a person would feel the same downward pull of gravity. Einstein asserted that these effects were actually the same. A far cry from Newton's view of gravity as a force acting at a distance!

GRAVITATIONAL TIME DILATION

Einstein's Special Theory of Relativity predicted that time does not flow at a fixed rate: moving clocks appear to tick more slowly relative to their stationary counterparts. But this effect only becomes really significant at very high velocities that app roach the speed of light.

When "generalized" to include gravitation, the equations of relativity predict that gravity, or the curvature of spacetime by matter, not only stretches or shrinks distances (depending on their direction with respect to the gravitational field) but also w ill appear to slow down or "dilate" the flow of time.

In most circumstances in the universe, such time dilation is miniscule, but it can become very significant when spacetime is curved by a massive object such as a black hole. For example, an observer far from a black hole would observe time passing extremely slowly for an astronaut falling through the hole's boundary. In fact, the distant observer would never see the hapless victim actually fall in. His or her time, as measured by the observer, would appear to stand still. The slowing of time near a very simple black hole has been simulated on supercomputers at NCSA and visualized in a computer-generated animation.

GRAPPLING WITH RELATIVITY

In the decade after its publication in 1916, Einstein's Theory of General Relativity led to a burst of experimental activity in which many of its predictions were vindicated. These predictions were encapsulated in a series of field equations that laid the foundation for all subsequent research into relativity and partly for modern cosmology as well.

The Math Behind Einstein's Vision

The mathematics behind the Einstein Field Equations not only presented a formidable challenge to solve, but also led to seemingly bizarre consequences, particularly those of black holes and gravitatio nal waves. At the time they were postulated, both were dismissed by many experts as mathematical aberrations. It remains to be seen whether either truly exist.

THE EINSTEIN FIELD EQUATIONS

INTRODUCTION

When expressed mathematically, Einstein's gravitation theory yields a set of formulas known collectively as the Einstein Field Equations. In a nutshell, this equation describes the properties of a gravitational field surroun ding a given mass. Stated simply, the equation describes how an object curves space and how the curvature, in turn, stretches or squeezes matter in three spatial directions: north-south, east-west and up-down.

However, applying the equations is no easy task. Mathematically, General Relativity is built upon ten so-called "coupled hyperbolic-elliptic nonlinear partial differential equations" which take many pages to write down (and a deep br eath just to say!). For a mathematician, the difficulties lie in the fact that the equations are nonlinear and coupled also and in their sheer number of terms.

After he published his famous paper in 1916, Einstein later conceded that the mathematical difficulties of his General Theory of Relativity were a "very serious" impediment to its further development. So serious, in fact, that it took nearly 75 years before the best minds in the field could come close to solving the equations stated by the theo ry. Now, high performance computers permit more accurate modeling of the distortions of spacetime by massive objects, including black holes.

Physicists are motivated to grapple with the complexity of Einstein's field equations because they can be used, theoretically at least, to describe all possible spacetime scenarios, from the collisions of black holes to the gravit ational interactions of irregular chunks of matter. In practice, researchers had to first restrict their calcuations to very simple systems in order to manage the equations.

THE SYMBOLIC FORM

This elegant symbolic formulation of Einstein's general theory of relativity cannot be used for actual calculations, but it clearly shows the principle that "matter tells spacetime how to curve, and curved space tells matter how to move"(John Wheeler, Princeton University and the University of Texas at Austin) . The left side of the equation contains all the information about how space is curved, and the right side contains all the information about the location and motion of the matter. General relativity is beautiful and simple (to a physicist), but mathemat ically it's very complicated and subtle.

Spherical Symmetry

The first person to apply Einstein's general theory was the German astrophysicist Karl Schwarzschild. It was 1916, and the First World War was still raging. Just a few days after reading Einstein's newly published theory while stationed at the Russian front, Schwarzschild began to figure out its consequences for the gravitational fields of stars.

For his first attempt, Schwarzschild simplified the problem by considering a perfectly spherical star at rest, ignoring the effects of the star's interior. He sent his preliminary solution of the star's spacetime curvature to Einstein, who reported the results on Schwarzschild's behalf at a physics meeting in January. The curva ture of spacetime predicted by the solution became known as the Schwarzschild geometry and had profound implications on future research into gravitation and cosmology.

A few weeks later, Schwarzschild sent a second paper, this time describing the spacetime curvature inside a star. Tragically, Schwarzschild died a few months later of an illness he contracted while at the Russian front.

Schwarzschild was describing a singularity, a region of infinite spacetime curvature that is postulated to lie within what has more recently been termed a black hole. Einstein considered the "Schwarzschild singularity" and black holes as biz zarre constructs, resisting the logic of his own theory right up to his death in 1955.

However, though debate continues on the nature of singularities, since the 1960s there has been mounting indirect evidence that black holes might exist in places where, for example, a collapsed star's intense gravitational field allows nothing, including light, to escape. In that sense, the star disappears from the visible universe and forms what is now called a black hole.

Had Schwarzschild lived, he would have likely developed more elaborate scenarios of spacetime curvature. Nevertheless, his early achievement was not topped for nearly 50 years, when the simple, spherical geometry of his solution was finally expanded to co nsider the gravitational effects of a spinning star.

Now, in the 1990s, astrophysicists are using supercomputers to extend the calculations to more complex spacetime geometries, including spinning objects that no longer retain their simple, spherical symmetry. Such objects typically exhibit "axisymmetry" – symmetry about one axis (like a football) – or no symmetry at all. The mathematics, however, becomes very difficult to state, let alone solve analytically.

If one considers an axisymmetric object, one that exhibits symmetry about a single axis, the equations expand dramatically into dozens of pages. At this point, the equations are so complex they can only be manipulated reliably by special computer software. Advanced computational techniques are needed to study distorted, spinning black holes, or the head-on collision of two axisymmetric black holes. This exhibit presents the results of many such black hole calculations.

FULL GENERALITY

In three-dimensional, time-dependent spacetimes, the lack of symmetry results in yet more variables. Just for the gravitational field itself, there are at least 16 variables in the standard formulation (developed for computation by Arrow itt, Deser and Misner in 1962). These variables depend on further coordinates, three for space and one for time, so there are many, many more terms in the equations.

Fortunately, it is possible to apply the General Theory of Relativity to the "real world" without having to completely solve the mathematics. For one thing, we can try to find examples in the universe where objects or processes exhibit the behaviours that the theory predicts. In fact, several observational tests of the General Theory of Relativity were proposed by Einstein and others almost immediately after its publication in 1916.

PUTTING RELATIVITY TO THE TEST

Though Einstein based his theory of gravitation on deep theoretical principles, he and others proposed a number of experimental tests of the theory soon after its publication.

Bending Light

The first prediction put to test was the apparent bending of light as it passes near a massive body. This effect was conclusively observed during the solar eclipse of 1919, when the Sun was silhouetted against the Hyades star cluster, for which the positi ons were well known.

Sir Arthur Eddington stationed himself on an island off the western coast of Africa and sent another group of British scientists to Brazil. Their measurements of several of the stars in the cluster showed that the light from these stars was indeed bent as it grazed the Sun, by the exact amount of Einstein's predictions. Einstein became a celebrity overnight when the results were announced.

The apparent displacement of light results from the warping of space in the vicinity of the massive object through which light travels. The light never changes course, but merely follows the curvature of space. Astronomers now refer to this displacement o f light as gravitational lensing.

But the Sun's gravity is relatively weak compared with what's out there in the depths of space. In the dramatic example of gravitational lensing below, the light from a quasar (a young, distant galaxy that emits prodigious amounts of radio energy) 8 billi on light years away is bent round by the gravity of a closer galaxy that's "only" 400 million light years distant from Earth.

The Einstein Cross

Four images of the quasar appear around the central glow formed by the nearby galaxy. The Einstein Cross is only visible from the southern hemisphere.

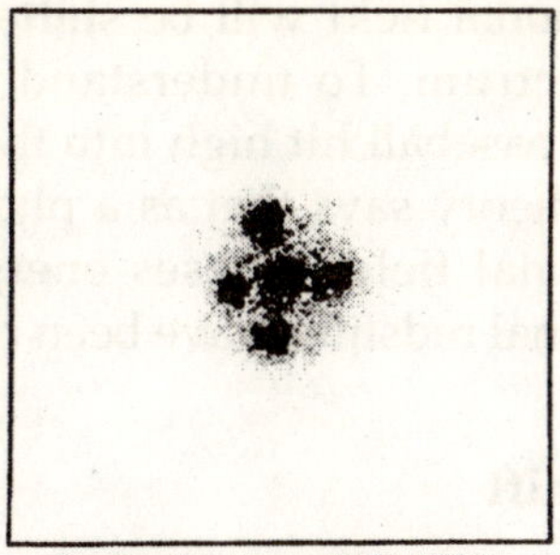

Gravitational Lens G2237+0305

Mercury's Changing Orbit

In a second test, the theory explained slight alterations in Mercury's orbit around the Sun.

Since almost two centuries earlier astronomers had been aware of a small flaw in Mercury's orbit around the Sun, as predicted by Newton's laws. As the closest planet to the Sun, Mercury orbits a region in the solar system where spacetime is disturbed by t he Sun's mass. Mercury's elliptical path around the Sun shifts slightly with each orbit such that its closest point to the Sun (or "perihelion") shifts forward with each pass. Newton's theory had predicted an advance only half as large as the one actually observed. Einstein's predictions exactly matched the observation.

Since almost two centuries earlier astronomers had been aware of a small flaw in Mercury's orbit around the Sun, as predicted by Newton's laws. As the closest planet to the Sun, Mercury orbits a region in the solar system where spacetime is disturbed by the Sun's mass. Mercury's elliptical path around the Sun shifts slightly with each orbit such that its closest point to the Sun (or "perihelion") shifts forward with each pass. Newton's theory had predicted an advance only half as large as the one actually observed. Einstein's predictions exactly matched the observation.

GRAVITATIONAL REDSHIFT

According to General Relativity, the wavelength of light (or any other form of electromagnetic radiation) passing through a gravitational field will be shifted towards redder regions of the spectrum. To understand this gravitational redshift, think of a baseball hit high into the air, slowing as it climbs. Einstein's theory says that as a photon fights its way out of a gravitational field, it loses energy and its colour reddens. Gravitational redshifts have been observed in diverse settings.

Earthbound Redshift

In 1960, Robert V. Pound and Glen A. Rebka demonstrated that a beam of very high energy gamma rays was ever so slightly redshifted as it climbed out of Earth's gravity and up an elevator shaft in the Jefferson Tower physics building at Harvard University. The redshift predicted by Einstein's Field Equations for the 74 ft. tall tower was but two parts in a thousand trillion. The gravitational redshift detected came within ten per cent of the computed value.

Solar Redshift

In the 1960s, a team at Princeton University measured the redshift of sunlight. Though small, given the Sun's mass and density, the redshift matched Einstein's prediction very closely.

White Dwarf Redshift

Take a star like the white dwarf star, Sirius B that is 61,000 times denser than the Sun. Its gravitational field is correspondingly much stronger and so is the redshift for the light it emits: 30 times greater, ac cording to earlier observations from the Mount Wilson Observatory taken by W.S. Adams way back in 1924. Still larger redshifts have more recently been detected in studies of so-called neutron stars — collapsed stars that are even denser. What about the redshift caused by a black hole? It can be thought of as infinite. In other words, photons inside the hole are so redshifted they can never get out!

BLACK HOLES AND BEYOND

Einstein's general theory of relativity describes gravity as a curvature of spacetime caused by the presence of matter. If the curvature is fairly weak, Newton's laws of gravity can explain most of what is observed. For example, the regular motions of the planets. Very massive or dense objects generate much stronger gravity. The most compact objects imaginable are predicted by General Relativity to have such strong gravity that nothing, not even light, can escape their grip. Black holea re probably the most researched and best known of the topics dealing with relativity. Their existence has been speculated for over 200 years, but has yet to be proven.

A black hole is a region of space where gravity is so strong that nothing, not even light, can escape. (According to relativity, nothing can travel faster than light, so this makes sense.)

But what would cause this strong of a gravitational force? According astronomer John Michell in 1783, it would have to be the collapse of a massive star. And today we know that these massive stars exist-neutron stars.

A neutron star is the final phase of a star more than two times as large as the sun. The sun would collapse into a white dwarf-a dense, earth-sized ball of gas that can collapse no further due to the outward pressure of electrons spinning at near the speed of light.

But in a large star, the gravitational force is so strong that the electrons are pulled into the atomic nucleus where they combine with protons to form neutrons. A neutron star is small and unchanged-only a few miles in diameter. Then one of two things can happen.

That 'strong force'-the force that holds the atomic nucleus together can abruptly stop the implosion and cause the start to burst into a supernova. But if they star is more than two times the sun's size, nothing can stop its collapse into a small, dense object that traps even light- a black hole.

How does aone picture this in the rubber sheet model? Imagine placing an object so massive on the sheet that it cause its indentation to pinch off. This would cause the rest of the sheet to angle down towards the hole, and smaller objects would roll in towards this huge 'gravitational pull'.

Scientists today call such an object a black hole. Why black? Though the history of the term is interesting, the main reason is that no light can escape from inside a black hole: it has, in effect, disappeared from the visible universe.

Do black holes actually exist? Most physicists believe they do, basing their views on a growing body of observations. In fact, present theories of how the cosmos began rest in part on Einstein's work and predict the existence of both singularities and the black holes that contain them. Yet Einstein himself vigorously denied their reality, believing, as did most of his contemporaries, that black holes were a mere mathematical curiosity. He died in 1955, before the term "black hole" was coined or understood and observational evidence for black holes began to mount.

Black holes are thought to form from stars or other massive objects if and when they collapse from their own gravity to form an object whose density is infinite: in other words, a singularity. During most of a star's lifetime, nuclear fusion in the core generates electromagnetic radiation, including photons, the particles of light. This radiation exerts an outward pressure that exactly balances the inward pull of gravity caused by the star's mass.

As the nuclear fuel is exhausted, the outward forces of radiation diminish, allowing the gravitation to compress the star inward. The contraction of the core causes its temperature to rise and allows remaining nuclear material to be used as fuel. The star is saved from further collapse — but only for a while.

Eventually, all possible nuclear fuel is used up and the core collapses. How far it collapses, into what kind of object, and at what rate, is determined by the star's final mass and the remaining outward pressure that the burnt-up nuclear residue (largely iron) can muster. If the star is sufficiently massive or compressible, it may collapse to a black hole. If it is less massive or made of stiffer material, its fate is different: it may become a white dwarf or a neutron star.

As was mentioned earlies, black holes are not known for sure to exist. But there may be one at the centre of our galaxy, or in the binary system Cygnus X-1, 6000 light years from Earth.

Besides looking for the obvious area of 'nothingness' in space, there are other clues to the presence of one. In a binary system, one looks for accretion disks, a disk of gas that swirls into the suspected black hole from a neighbouring star.

There is also the gravitational red shift. Longer light waves (red colours) have less energy. Photons of light will lose more and more energy as they try to escape the steadily stronger gravity of a black hole. The wavelengths lengthens until we must use radio and infrared telescopes to detect it, and then it either becomes to long to detect or it disappears.

The newest proof of black holes may be quasars-stars that radiate enough energy like an entire galaxy. These could be 'white holes', the opposite end of a black hole.

Black holes may have another way of being detected - Hawking Radiatin. Although nothing escapes from a black hole, Steven Hawking theorized that something must, because black holes have entropy. Anything with entropy must have a temperature, and anything with a temperature must emit radiation. The empty space around a black hole, he says, contains miniscule virtual particles that have positive and negative charges. Normally they combine to annhiliate each other. Sometimes the negative one will fall into the black hole, and the positively charged one will escape.

We have, been, plenty of proof that black holes exist. We just have yet to prove that they do exist. If the do, the all important question would be;

ANATOMY OF A BLACK HOLE

By definition a black hole is a region where matter collapses to infinite density, and where, as a result, the curvature of spacetime is extreme. Moreover, the intense gravitational field of the black hole prevents any light or other electromagnetic radiation from escaping. But where lies the "point of no return" at which any matter or energy is doomed to disappear from the visible universe?

Although black holes would seem the upitome of chaos, they have a definite structure, and it is even thought that the more matter and energy they pull in, the more ordered they become. (This distintly goes against Newton's Laws of Physics, but who's going to tell the black hole that?)

The outermost edge of a black hole is the event horizon, or Schwarzschild radius. It's the point at which the star collapsed into a black hole, and its size is therefore proportional to the star's mass.

That's the only definite part of a black hole. It used to be also generally agreed that there is a singularity at the centre of a black holes where the laws of physics break down, but recently, Stephen Hawking, the leading expert on black holes, has begun to suspect that this may not be true.

Although black holes have the same structure, they are not all alike. Size varies greatly among black holes, with smaller ones causing more distortion of space-time; space-time would have to wrap itself more tightly around a smaller one to close it off.

The physical difference with the most impact, though, is whether it is a rotating or non-rotating black hole. A still black hole is just as was discribed: event horizon, that singularity. But a rotating black hole has two even horizons, and makes some major differences.

Inside the first event horizon, it is just like a non-rotating black hole (the rotation, by the way, is from the rotation of the star before it collapses): gravity pulling you in faster and faster.

But when you hit the second event horizon, it's like the eye of a storm: There is no singularity, and space-time seems to return to normal.

There is a problem, of course-there may be anti-gravity after the second event horizon, so you would be spit right back out. To where? Perhaps an alternate universe.

The Event Horizon

Applying the Einstein Field Equations to collapsing stars, German astrophysicist Kurt Schwarzschild deduced the critical radius for a given mass at which matter would collapse into an infinitely dense state known as a singularity. For a black hole whose mass equals 10 suns, this radius is about 30 kilometers or 19 miles, which translates into a critical circumference of 189 kilometers or 118 miles.

Schwarzschild Black Hole

If you envision the simplest three-dimensional geometry for a black hole, that is a sphere (known as a Schwarzschild black hole), the black hole's surface is known as the event horizon. Behind this horizon, the inward pull of gravity is overwhelming and no information about the black hole's interior can escape to the outer universe.

Apparent versus Event Horizon

As a doomed star reaches its critical circumference, an "apparent" event horizon forms suddenly. Why "apparent?" Because it separates light rays that are trapped inside a black hole from those that can move away from it. However, some light rays that are moving away at a given instant of time may find themselves trapped later if more matter or energy falls into the black hole, increasing its gravitational pull. The event horizon is traced out by "critical" light rays that will never escape or fall in.

Even before the star meets its final doom, the event horizon forms at the centre, balloons out and breaks through the star's surface at the very moment it shrinks through the critical circumference. At this point in time, the apparent and event horizons merge as one: the horizon. For more details, see the caption for the above diagram.

The distinction between apparent horizon and event horizon may seem subtle, even obscure. Nevertheless the difference becomes important in computer simulations of how black holes form and evolve.

Beyond the event horizon, nothing, not even light, can escape. So the event horizon acts as a kind of "surface" or "skin" beyond which we can venture but cannot see. Imagine what happens as you approach the horizon, then cross the threshold.

The Singularity

At the centre of a black hole lies the singularity, where matter is crushed to infinite density, the pull of gravity is infinitely strong, and spacetime has infinite curvature. Here it's no longer meaningful to speak of space and time, much less spacetime. Jumbled up at the singularity, space and time cease to exist as we know them.

THE LIMITS OF PHYSICAL LAW

Newton and Einstein may have looked at the universe very differently, but they would have agreed on one thing: all physical laws are inherently bound up with a coherent fabric of space and time.

At the singularity, though, the laws of physics, including General Relativity, break down. Enter the strange world of quantum gravity. In this bizzare realm in which space and time are broken apart, cause and effect cannot be unraveled. Even today, there is no satisfactory theory for what happens at and beyond the singularity.